FM 3-05.202 (FM 31-20-3)

Special Forces Foreign Internal Defense Operations

February 2007

Headquarters, Department of the Army

FM 305.202

Special Forces Foreign Internal Defense Operations

February 2007

FM 3-05.202
Special Forces
Foreign Internal Defense Operations

ISBN-13: 978-1481835916

ISBN-10: 1481835912

Proudly Printed in the
U.S.A

Field Manual
No. 3-05.202 (31-20-3)

Headquarters
Department of the Army
Washington, DC, 2 February 2007

Special Forces Foreign Internal Defense Operations

Contents

*This publication supersedes FM 31-20-3, 20 September 1994.

Figures

Preface

Field manual (FM) 3-05.202, *Special Forces Foreign Internal Defense Operations*, supports FM 3-05.20, *(C) Special Forces Operations (U)*, which is the keystone manual of Special Forces (SF). FM 3-05.202 defines the current United States (U.S.) Army SF concept of planning and conducting SF foreign internal defense (FID) missions.

PURPOSE

As with all doctrinal manuals, FM 3-05.202 is authoritative but not directive. It serves as a guide and does not preclude SF units from developing their own standing operating procedures (SOPs) to meet their needs. It explains planning, roles of SF in FID, and the various programs that SF Soldiers participate in to conduct FID operations. Other SF primary missions are discussed at length in appropriate manuals in the series.

SCOPE

The primary users of this manual are commanders, staff officers, and operational personnel at the team (Special Forces operational detachment A [SFODA]), company (Special Forces operational detachment B [SFODB]), and battalion levels (Special Forces operational detachment C [SFODC]). This FM is specifically for SF; however, it is also intended for use Armywide to improve the integration of SF into the plans and operations of other special operations forces (SOF) and conventional forces.

APPLICABILITY

Commanders and trainers should use this and other related manuals in conjunction with command guidance, the Army Training and Evaluation Program (ARTEP), and the mission training plan (MTP) to plan and conduct successful FID operations. This publication applies to the Active Army, the Army National Guard (ARNG)/Army National Guard of the United States, and the United States Army Reserve (USAR) unless otherwise stated.

ADMINISTRATIVE INFORMATION

The proponent of this manual is the United States Army John F. Kennedy Special Warfare Center and School (USAJFKSWCS). Submit comments and recommended changes to Commander, USAJFKSWCS, ATTN: AOJK-DTD-SF, Fort Bragg, NC 28310-5000.

Unless this publication states otherwise, masculine nouns and pronouns do not refer exclusively to men.

This page intentionally left blank.

Chapter 1

The Nature of Foreign Internal Defense

FID is a joint, multinational, and interagency effort. SOF, particularly SF and Psychological Operations (PSYOP) and Civil Affairs (CA) forces are well suited to conduct or support FID operations because these forces have unique functional skills and cultural and language training. FID is a legislatively directed activity for SOF (although it is not exclusively a SOF mission) under the 1986 Goldwater-Nichols Department of Defense Reorganization Act. SOF may conduct FID unilaterally in the absence of any other military effort, support other ongoing military or civilian assistance efforts, or support the employment of conventional forces. In the National Security Strategy (NSS) of the United States (2006), the strategy states that "Regional conflicts can arise from a wide variety of causes, including poor governance, external aggression, competing claims, internal revolt, tribal rivalries, and ethnic or religious hatreds." U.S. policy currently deals with these threats through the indirect use of military force in concert with the diplomatic, informational, and economic elements of national power. Direct use of military force is the exception rather than the rule. This approach relies on supporting the efforts of the government of the nation in which the problem is developing.

Let every nation know, whether it wishes us well or ill, that we shall pay any price, bear any burden, meet any hardship, support any friend, oppose any foe to assure the survival and the success of liberty.

President John F. Kennedy
Inaugural Address, January 20, 1961

OVERVIEW

1-1. Nations in time of need often look to other nations to provide assistance. These nations seeking assistance are often struggling to quell unrest within their borders or are seeking ways to strengthen or further professionalism within their military. Internal problems or potential problems could stem from economic issues, a populace dissatisfied with the government, social unrest, or terrorism. The United States has historically promoted democracy and freedom in other nations by assisting nations seeking solutions to improve security and unrest within its borders. Numerous U.S. organizations, civilian and military, support this effort. For the military, this effort is FID. Joint Publication (JP) 1-02, *Department of Defense Dictionary of Military and Associated Terms*, defines FID as the "participation by civilian and military agencies of a government in any of the action programs taken by another government or other designated organization to free and protect its society from subversion, lawlessness, and insurgency."

1-2. FID planners must consider all the elements of national power, to include diplomatic, informational, military, and economic. The National Security Council (NSC) is responsible for planning guidance for FID at the strategic level. The Department of State (DOS) is normally designated the lead agency for execution of FID programs. However, military assistance is often required to provide a secure environment to accomplish a host nation's (HN's) goals. The Department of Defense (DOD) provides personnel and equipment to help achieve FID objectives.

1-3. Supporting the FID requirements and identified needs of an HN is the compilation of the national military strategy (NMS), joint plans, and the geographic combatant commander's (GCC's) developed plans and integrated military activities. These plans are based on U.S. policies developed with friends, allies, and partner nations. These strategic commitments with various nations may lead to their enhanced security, greater cooperation, and stronger worldwide alliances. Commitments to other nations based on providing a more secure environment lead to various programs to help build or enhance their internal defense and development (IDAD) program or provide assistance in other areas. Military involvement in FID activities could range from training HN forces to secure a port waterway to providing courses to combat terrorism. FID could also be interrelated with other military operations such as unconventional warfare (UW) or actual combat operations. One unit could have a FID mission to train a force while another military unit works with that trained force and conducts actual combat operations.

1-4. The strategic end state is an HN capable of successfully integrating military force with other instruments of national power to eradicate lawlessness, insurgency, subversion, and terrorism. Ultimately, FID efforts are successful if they preclude the need to deploy large numbers of U.S. military personnel and equipment. Types of military operations related to FID are nation assistance (NA) and/or support to counterinsurgency (COIN); counterterrorism (CT); peace operations (PO); DOS support to counterdrug (CD) operations; and foreign humanitarian assistance (FHA). These categories may, to some degree, include FID operations as an integral component in supporting the fight against subversion, lawlessness, insurgency, and terrorism. FID programs are distinct and will vary from country to country to support that country's IDAD program.

INTERNAL DEFENSE AND DEVELOPMENT

1-5. IDAD is the full range of measures taken by a nation to promote its growth and protect itself from subversion, lawlessness, and insurgency. It focuses on building viable institutions (political, economic, military, and social) that respond to the needs of the society. IDAD is the HN's program. The HN has responsibility and control of the program. Development programs that are carefully planned and implemented and properly publicized can serve the interests of population groups and deny exploitable issues to the insurgents. Security programs provide an atmosphere of peace within which development can take place.

1-6. The IDAD strategy is founded on the assumption that the HN is responsible for the development and execution of its own programs to prevent or defeat subversion, lawlessness, and insurgency. The fundamental thrust of the IDAD strategy is toward preventing the escalation of internal conflict. Anticipating and defeating the threat posed by specific organizations and working to correct conditions that prompt violence are effective means of prevention. If subversion, lawlessness, insurgency, or terrorism occurs, emphasis is placed on holding down the level of violence. The population must be mobilized to participate in IDAD efforts. Thus, IDAD is an overall strategy for the prevention of these activities and, if an insurgency or terrorism should develop, for COIN and CT activities. U.S. Army FID operations contribute to the overall IDAD strategy of the HN and are based on integrated military and civilian programs.

UNITED STATES NATIONAL OBJECTIVES AND POLICY

1-7. A basic premise of U.S. foreign policy is that the security of the United States and its fundamental values and institutions will be best preserved and enhanced as part of a community of free and independent nations. In this regard, the United States endeavors to encourage other countries to do their part in the preservation of this freedom and independence. The objective is to support U.S. interests by means of a common effort. This common effort makes use of instruments of national power to support an HN. The diplomatic instrument is often first used to show U.S. commitment. The political system within the HN is key in providing the stability and must be willing to improve the stability within its borders. The economic instrument has influence across all aspects of FID. (Figure 1-1, page 1-3, shows the FID framework.) In many cases, FID is incorporated into HN programs within nations that are usually less developed and require means to improve the economy. HN programs can range from favorable trade arrangements to military financing. The informational instrument gets the message out to the public. Information operations (IO) portray the positive efforts and accomplishments of the HN. These operations also publicize the U.S. support to the HN and U.S. efforts to improve the HN. Although the focus of this publication is on the military instrument, the military instrument is primarily a supporting role to the overall FID program. This military instrument provides support in the following three ways:

- *Indirect support.* Indirect support builds strong national infrastructures through economic and military capabilities that contribute to self-sufficiency. This can include unit exchange programs, personnel exchange programs (PEPs), individual exchange programs, and combination programs.
- *Direct support.* In direct support, U.S. forces provide direct assistance to the HN civilian populace or military. This support can be evaluation, training, limited information exchange, and equipment support.
- *Combat operations.* The President must approve combat operations. Combat operations are a temporary solution until HN forces can stabilize the situation and provide security for the populace. Emphasis should be placed on HN forces in the forefront during these operations to maintain HN legitimacy with the population. Combat operations can include COIN operations.

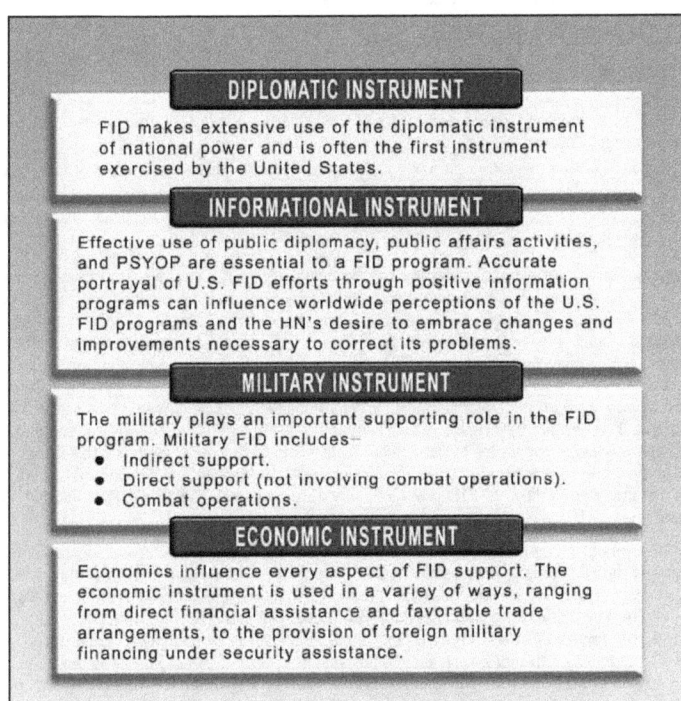

Figure 1-1. The FID framework

1-8. Those governments that lack the will to address their social, economic, or political problems are unlikely to benefit from outside assistance. However, governments that do mobilize their human and material resources may find that outside help, to include U.S. security assistance (SA), makes a critical difference. Where significant U.S. national interests are involved, the United States may provide economic and military assistance to supplement the efforts of such governments.

1-9. The creation of a relatively stable internal environment, one in which economic growth can occur and the people are able to determine their own form of government, is a primary U.S. objective. Economic assistance, either supplied by the United States through bilateral agreements or by several nations through multilateral agreements, may help achieve this objective.

1-10. The primary responsibility for creating a stable atmosphere through the commitment and use of all its internal resources rests with the threatened government. Under certain conditions, U.S. policy supports supplementing local efforts to maintain this order and stability. These conditions are as follows:

- The internal disorder is of such a nature as to pose a significant threat to U.S. national interests.
- The threatened country is capable of effectively using U.S. assistance.
- The threatened country requests U.S. assistance.

1-11. The United States Government (USG) spends billions of dollars a year, with certain expectations, in programs to improve allied and friendly nations. There are numerous benefits for the U.S. military to conduct FID throughout the world. These benefits include—

- FID programs help build and foster favorable relationships that promote U.S. interests. In many cases, these programs lead to the establishment of personal and unit relationships.
- FID programs strengthen friendly nation capabilities, which ultimately strengthen U.S. security concerns.
- Many of the foreign areas aided by the United States provide U.S. forces with peacetime and contingency access.
- Training exercises with foreign nations that increase the proficiency and skills of U.S. forces.
- Improvement of U.S. forces' regional knowledge of specific areas, which can be disseminated throughout the force (environment, terrain, social, political, economic, culture, and beliefs).
- Improved effectiveness of the War on Terrorism.

1-12. Subversion, lawlessness, and insurgency are the result of specific conditions within a nation. They may stem from the population's perception that they are suffering from conditions such as poverty, unemployment, religious disparity, political issues, crime, or tribal unrest. These conditions have historically set the stage for lawlessness and insurgent activity against an established government. This type of internal strife or conflict within a nation's borders may remain a local problem or expand, which allows an outside source to influence or create opposition toward the legitimate government. In some cases, outside sources may threaten the HN's stability by exploiting the conditions within that nation, to further their own cause. This outside influence may even establish itself within the HN to promote and support civil unrest. These types of conditions promote insurgencies and their violent solutions, like terrorism. U.S. military involvement in FID has traditionally focused on COIN. Although much of the FID effort remains focused on this important area, U.S. FID programs may aim at other threats to an HN's internal stability, such as terrorism.

1-13. Identification of the root cause of the problem, analysis of the environment, and identification of the specific needs of the HN are key in tailoring military support to assist an HN's IDAD program. Emphasis should be on helping the HN address the root cause of instability in a preventative manner rather than reacting to threats. The United States will support specific nations based on U.S. policy toward that nation or region and will implement FID programs to support that nation through GCC security cooperation programs. FID programs of all types, such as humanitarian assistance (HA) and CT programs, can prevent, reduce, or stop mitigating factors that can contribute to the beginning or spread of terrorism and insurgencies. FID activities implemented through the GCC may ultimately lead to stability within that nation or region and effectively reduce threats to the United States.

Chapter 2

United States Organization
for Foreign Internal Defense

To assist a country with its IDAD efforts, one must understand the political climate, social attitudes, economic conditions, religious considerations, philosophy or plan of the insurgents, the host government, and the local population. One should also understand how the United States implements diplomatic, economic, informational, and military instruments in a coordinated and balanced combination to help remedy the situation.

MISSIONS

2-1. FID is the role the U.S. military plays in the overall effort of the USG to help a nation free or protect its society from an existing or potential threat. U.S. FID operations work on the principle that it is the inherent responsibility of the threatened government to use its leadership and organizational and materiel resources to take the political, economic, and social actions necessary to defeat subversion, lawlessness, insurgency, and terrorism. The U.S. military can provide resources such as material, advisors, and trainers to support these FID operations. In instances where it is in the security interest of the United States, and at the request of the HN, more direct forms of U.S. military support may be provided, to include combat forces. The following principles apply to FID:

- All U.S. agencies involved in FID must coordinate with one another (Figure 2-1, page 2-2) to ensure that they are working toward a common objective and deriving optimum benefit from the limited resources applied to the effort.
- The U.S. military seeks to enhance the HN military and paramilitary forces' overall capability to perform their IDAD mission. An evaluation of the request and the demonstrated resolve of the HN government will determine the specific form and substance of U.S. assistance, as directed by the President.
- Specially trained, selected, and jointly staffed U.S. military survey teams, including intelligence personnel, may be made available. U.S. military units used in FID roles should be tailored to meet the conditions within the HN.
- U.S. military support to FID should focus on assisting HNs in anticipating, precluding, and countering threats or potential threats.

MILITARY SUPPORT

2-2. Emphasis on IDAD when organizing, planning, and executing military support to a FID program is essential. This emphasis helps the HN address the root causes of instability in a preventive manner rather than reacting to threats. COIN (Appendix A) has traditionally been the focus of U.S. military involvement in FID. Although much of the FID effort remains focused on this important area, U.S. FID programs may aim at other threats to the internal stability of the HN, such as civil disorder, illicit drug trafficking, and terrorism.

2-3. At the national level, the USG has two fundamental courses of action (COAs) to assist an ally against a potential or actual threat to its security:

- *Security assistance.* One COA is the application of a wide variety of programs executed by different USG agencies. These programs aid developing nations to make economic, political, humanitarian, and military improvements and are defined under the broad title of U.S. foreign assistance programs, humanitarian and civic assistance (HCA) programs, and SA programs. These programs can be a part of a nation's developed FID program.

- *Foreign internal defense.* The deployment of U.S. combat forces to assist an ally in internal defense is another COA. Assistance may occur during peacetime or conflict. The U.S. Army is assigned various missions in support of the national FID objectives. SF units may be required to perform FID missions ranging from preservation of a secure and stable environment to assisting an ally to defeat an internal threat through large-scale combat operations.

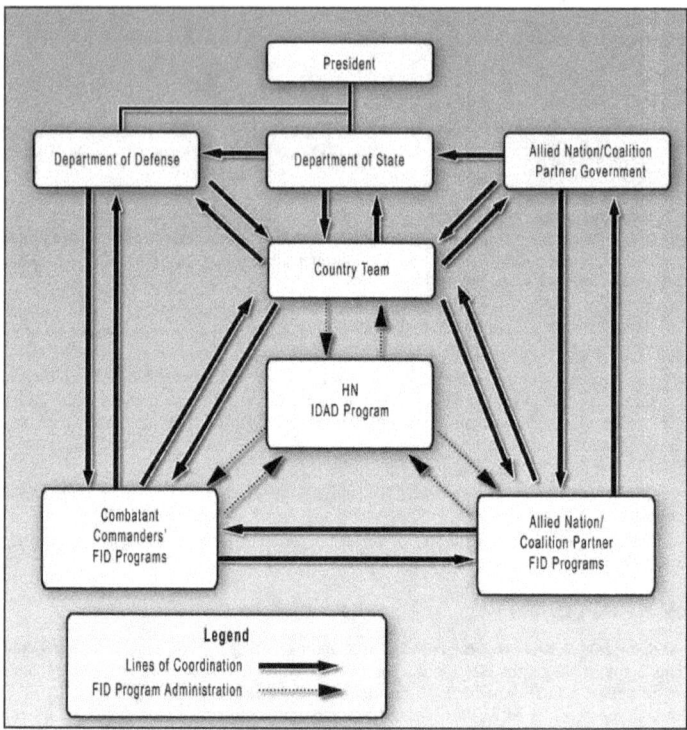

Figure 2-1. FID coordination

NATIONAL-LEVEL ORGANIZATIONS

2-4.   The United States uses national-level organizations in addressing IDAD issues. The following paragraphs discuss these national-level organizations.

NATIONAL SECURITY AGENCY

2-5.   The National Security Agency (NSA) was established by Presidential directive in 1952 to provide signals intelligence (SIGINT) and communications security activities for the government. Since then, the NSA has gained the responsibility for information systems security and operations security (OPSEC) training.

CENTRAL INTELLIGENCE AGENCY

2-6. The Central Intelligence Agency (CIA) is an independent agency, responsible to the President through the Director of National Intelligence (DNI), and accountable to the American people through the intelligence oversight committees of the U.S. Congress. The CIA's mission is to support the President, the NSC, and all officials who make and execute U.S. national security policy.

NATIONAL SECURITY COUNCIL

2-7. Created in 1947 by the National Security Act as amended in 1949, the NSC's formal members are the President, the Vice President, the Secretary of State, and the Secretary of Defense (SecDef). The director of the CIA, the head of the Joint Chiefs of Staff (JCS), the President's national security advisor (the assistant to the President for national security affairs, also director of the NSC), and the deputy advisor usually attend as invited guests. The council also has a civilian staff. The President appoints an executive secretary to head the staff.

DEPARTMENT OF STATE

2-8. The DOS is the federal department in the United States that sets and maintains foreign policies. The DOS is normally designated the lead agency for execution of FID programs and is overall responsible for the SA programs. The DOS is involved with policy formulation and execution of FID programs at the national level to the lowest levels within the HN.

BUREAU OF POLITICAL-MILITARY AFFAIRS

2-9. The Bureau of Political-Military Affairs, headed by an assistant secretary, is the principal link between DOS and DOD. This bureau provides policy direction in the areas of international security, SA, military operations, and defense trade. It is instrumental in the DOS's efforts to accomplish three major goals under the United States Strategic Plan for International Affairs—CT, regional stability, and HA.

COORDINATOR FOR INTERNATIONAL INFORMATION PROGRAMS

2-10. The coordinator for the Bureau of International Information Programs (BIIP) supports U.S. foreign policy objectives by influencing public attitudes in other nations. The coordinator for the BIIP also advises the President, his representatives abroad, and various departments and agencies on the implications of foreign opinion for present and contemplated U.S. policies, programs, and official statements. The BIIP uses various media and methods to—

- Publicize U.S. policies.
- Plan and conduct informative programs in support of U.S. or host government agencies.
- Counter propaganda hostile to U.S. interests.
- Coordinate U.S. overt PSYOP with guidance from the DOS.

UNITED STATES AGENCY FOR INTERNATIONAL DEVELOPMENT

2-11. The United States Agency for International Development (USAID) has the responsibility for carrying out nonmilitary U.S. foreign assistance programs and for the continuous supervision of all assistance programs under the Foreign Assistance Act of 1961. It is primarily concerned with developmental assistance and HCA. It also plans and implements overseas programs to improve economic and social conditions.

ARMS TRANSFER MANAGEMENT GROUP

2-12. The Arms Transfer Management Group is an interagency board that advises the Secretary of State on matters relating to SA program funding levels and arms transfer policies. The Under Secretary of State for Security Assistance, Science, and Technology chairs the Arms Transfer Management Group. The Group manages and coordinates weapons and equipment-related SA matters. The Group includes representatives from agencies throughout the executive branch who deal in SA matters. Its members may include, but are not limited to, the—

- NSC.
- DOD.
- Office of the Joint Chiefs of Staff (OJCS).
- CIA.
- Arms Control and Disarmament Agency.
- Office of Management and Budget.
- Department of Treasury.
- DOS.
- USAID.

2-13. The Group coordinates military assistance and military-related supporting assistance. This coordination encourages mutually supporting programs and increases the efficiency of the SA program.

BUREAU FOR INTERNATIONAL NARCOTICS AND LAW ENFORCEMENT AFFAIRS

2-14. The Bureau for International Narcotics and Law Enforcement Affairs (INL) advises the President, Secretary of State, other bureaus in the DOS, and other departments and agencies within the USG on the development of policies and programs to combat international narcotics and crime. A secretary who is under the direction of the Under Secretary for Political Affairs heads INL. INL programs support two of the DOS's strategic goals:

- To reduce the entry of illegal drugs into the United States.
- To minimize the impact of international crime on the United States and its citizens.

2-15. Counternarcotics and anticrime programs also complement the War on Terrorism, directly and indirectly, by promoting modernization of and supporting operations by foreign criminal justice systems and law enforcement agencies charged with the CT mission.

DEPARTMENT OF DEFENSE

2-16. Within the DOD, the Under Secretary of Defense for Policy (USD[P]) serves as the principal advisor and assistant to the SecDef for all matters concerned with the integration of DOD plans and policies with overall national security objectives. He also exercises direction, authority, and control over the Defense Security Cooperation Agency (DSCA). The DSCA is responsible for executing the following functions for the DOD:

- Administering and supervising SA planning and programs.
- Formulating and executing SA programs in coordination with other government programs.
- Conducting international logistics and sales negotiations with foreign countries.
- Managing the credit-enhancing program.
- Serving as the DOD focal point for liaison with U.S. industry concerning SA activities.

OFFICE OF THE JOINT CHIEFS OF STAFF

2-17. The OJCS plays a key role in the SA effort through the joint planning process. Key OJCS plans are the Joint Strategic Planning Document, the Joint Strategic Capabilities Plan (JSCP), and the Joint Intelligence Estimate for Planning. In addition, the OJCS continually reviews current and ongoing programs for specific countries and regions to ensure compatibility with U.S. global security interests.

GEOGRAPHIC COMBATANT COMMANDER

2-18. The GCCs integrate all military SA plans and activities with regional U.S. military plans. The role of the GCC is critical. His regional perspective is at the operational and strategic level of conflict. He identifies and applies military and certain humanitarian or civic action resources to achieve U.S. national strategic goals. With proper and timely employment, these resources minimize the likelihood of U.S. combat involvement.

UNITED STATES DIPLOMATIC REPRESENTATIVES TO A HOST NATION

2-19. U.S. organizations within an HN may be responsible for coordinating, planning, and resourcing numerous activities, to include FID. These organizations are composed of U.S. military and DOS personnel. The following describes the primary organizations within an HN involved with FID.

UNITED STATES DIPLOMATIC MISSIONS

2-20. The U.S. diplomatic mission to an HN includes representatives of all U.S. departments and agencies physically present in the country. The chief of mission (COM), normally an ambassador, ensures all in-country activities best serve U.S. interests as well as regional and international objectives. Two agencies that play an important role on the Country Team in supporting U.S. efforts to assist an HN in its IDAD efforts are the BIIP and the USAID.

COUNTRY TEAM

2-21. The Country Team is the point of coordination within the host country for the diplomatic mission. The members of the Country Team will vary depending on levels of coordination needed and the conditions within that country. It is usually headed by the chief of the U.S. diplomatic mission and composed of the senior member of each represented U.S. department or agency, as desired by the chief of the U.S. diplomatic mission. The purpose is to achieve a unity of effort, coordinate, and inform the various organizations of operations. Usually the primary military members are the defense attaché and the chief of the security assistance organization (SAO). Figure 2-2 shows the Country Team concept.

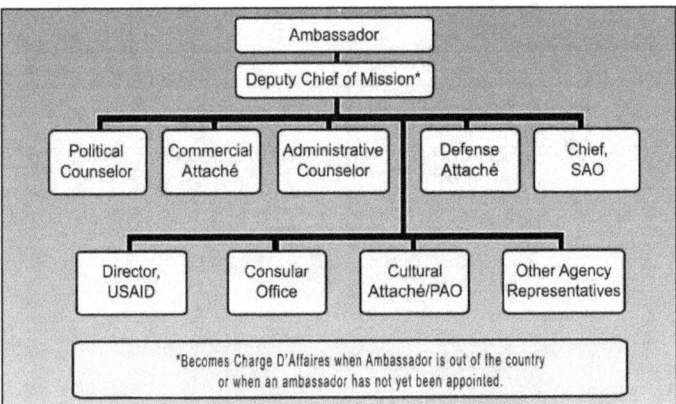

Figure 2-2. Country Team concept

SECURITY ASSISTANCE ORGANIZATION

2-22. The SAO is the in-country mechanism for ensuring that DOD SA management responsibilities, prescribed by law and executive direction, are properly executed. It oversees all foreign-based DOD elements with SA responsibilities. The SAO assists HN security forces by planning and administering military aspects of the SA program. SA offices also help the U.S. Country Team communicate HN assistance needs to policy and budget officials within the USG. The SAO may be known in-country by any number of personnel assigned, the functions performed, or the desires of the HN. Typical designations include Joint U.S. Military Advisory Group, Joint U.S. Military Group, U.S. Military Training Mission, Defense Field Office, or Office of Defense Cooperation. The Chief of the SAO reports to the theater GCC and is a member of the U.S. Embassy Country Team. Figure 2-3 shows the SAO departmental alignment. Figure 2-4 shows the SAO functional alignment.

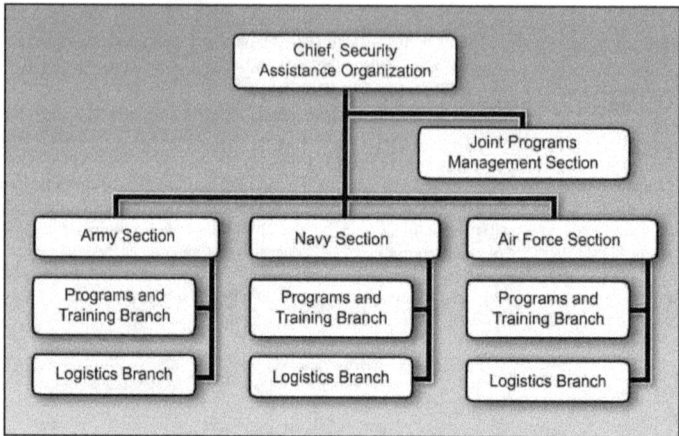

Figure 2-3. SAO departmental alignment

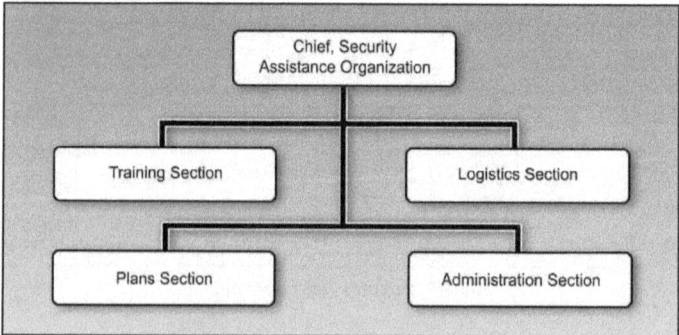

Figure 2-4. SAO functional alignment

UNITED STATES DEFENSE ATTACHÉ OFFICE

2-23. The United States Defense Attaché Office (DAO) performs representational functions on behalf of the SecDef, the Secretaries of the Military Services, the JCS, the Chiefs of the U.S. Military Services, and the GCC. The defense attaché serves as the military advisor to the COM, liaises with the HN military, and manages the U.S. SA and military-to-military programs. The DAO assists the GCC and his staff with FID programs by exchanging information on HN military, social, and political conditions.

UNITED STATES DEFENSE REPRESENTATIVE

2-24. The United States defense representative (USDR) represents the SecDef and the appropriate unified commanders for coordination of administrative and security matters for all DOD noncombatant command elements in the foreign country in which the USDR is assigned. The USDR in foreign countries is an additional duty title assigned to a military officer serving in a specifically designated position with prescribed authorities and functions. The USDR is the COM's single point of contact (POC) to assist the COM in carrying out his responsibilities. The responsibility of the USDR is established for U.S. Governmental administrative and security coordination only. USDR duties shall be performed in coordination with the respective GCC with geographic area responsibility.

MILITARY FORCES

2-25. In most instances, the application of U.S. military resources in support of an HN's IDAD programs will function through the framework of the organizations mentioned above. However, it may be necessary to expand U.S. assistance by introducing selected U.S. military forces. A joint task force (JTF) will normally be established to coordinate this effort. This JTF will—

- Exercise operational control (OPCON) of assigned U.S. military forces.
- Plan and conduct joint and combined exercises in coordination with the armed forces of the host government.
- Execute area command responsibilities for U.S. forces to ensure unity of effort.
- Specify the chain of command. However, units may be required to report to various organizations, to include DOS.

This page intentionally left blank.

Chapter 3

Planning

The 2005 National Defense Strategy of the United States of America provides that one of the United States military's most effective tools in prosecuting the War on Terrorism is to help train indigenous forces. As such, civilian and military agencies must assess what programs to conduct and plan the resources needed to ensure the programs succeed.

PLANNING OVERVIEW

3-1.   When an operational detachment conducts a FID mission in a foreign country, many levels of policy and planning will take place before their departure from the U.S. The specific mission the detachment will conduct can range from participating in a combined exercise to training an HN force on basic infantry skills. FID missions will fall under two major categories—those under the responsibility of DOD and those under DOS. To the detachment in the HN, the category may seem irrelevant; however, the activity or program the detachment has been deployed to participate in is governed by specific rules, funding, and conditions, depending on if the program falls under DOD or DOS oversight. The majority of the DOD and DOS activities are incorporated into the theater planning process. Through the theater planning process, identified activities are intended to help shape the theater in which the activities will be conducted. Depending on whether the mission has originated through DOD or DOS, how, where, and at what level the planning, coordination, and resourcing takes place will vary. For example, Title 22, United States Code (USC) governs DOS programs and indicates participants in these programs are noncombatants. Programs under Title 10, United States Code (10 USC) authorities do not restrict participants from being noncombatants.

DEPARTMENT OF DEFENSE GUIDANCE AND PLANNING

3-2.   Guidance produced from DOD ensures the force is focused on supporting the policy set forth from the President. The goal of a portion of this guidance is to accomplish security cooperation objectives without sacrificing combat readiness. The following produce guidance for security cooperation and ultimately lead to military FID operations.

NATIONAL MILITARY STRATEGY

3-3.   The NMS is the art and science of distributing and applying military power to attain national objectives in peace and war. This document articulates how the United States will employ the military element of power to support the national security objectives found in the President's NSS.

JOINT STRATEGIC PLANNING SYSTEM

3-4.   As the principal military advisor to the President and the SecDef, the Chairman of the Joint Chiefs of Staff (CJCS) shoulders a significant portion of the responsibility to develop strategic direction, strategic plans, and resource requirements for the national defense. The Joint Strategic Planning System (JSPS), supported by the joint warfighting capabilities assessment (JWCA) process, is the planning system used by CJCS to achieve these objectives. The JSPS process assists the CJCS with preparation of strategic plans; preparation and review of contingency plans; advice to the President and SecDef on requirements, programs, and budgets; and provision of net assessments on the capabilities of the Armed Forces of the

United States and its allies as compared with those of their potential adversaries. The JSCP is one of the products of the JSPS.

JOINT STRATEGIC CAPABILITIES PLAN

3-5. The JSCP provides guidance to the GCC and Service chiefs for accomplishing military tasks and missions based on current military capabilities. It also directs them to develop plans to support the strategy contained in the NMS and counter the threat using current military capabilities. It apportions resources to GCCs according to military capabilities resulting from completed program and budget actions and intelligence assessments. The capabilities of available forces, intelligence information, and guidance issued by the SecDef determine the resources apportioned. The JSCP directs the development of contingency plans to support national security objectives by assigning planning tasks and apportioning major combat forces and strategic lift capability to the GCCs. As a capabilities planning document, it represents the last phase of resource management. The JSCP apportions the resources provided by the Planning, Programming, and Budgeting System (PPBS) to develop operation plans (OPLANs). It provides guidance, missions, and resources to GCCs to develop concept plans (CONPLANs) and OPLANs to support FID missions. The JSCP provides a coherent framework for capabilities-based military advice provided to the President and SecDef.

JOINT OPERATION PLANNING AND EXECUTION SYSTEM

3-6. The Joint Operation Planning and Execution System (JOPES) provides the foundation for conventional command and control (C2) by national- and combatant command-level commanders and their staffs. It is designed to satisfy their information needs in the conduct of joint planning and operations. It includes joint operation planning policies, procedures, and reporting structures supported by communications and automated data processing systems. The JOPES is used to monitor, plan, and execute mobilization, deployment, employment, sustainment, and redeployment activities associated with joint operations. The JOPES is used in joint operational planning in either deliberate or crisis action procedures to meet the tasks identified in the JSCP.

3-7. SF planning at lower levels will use the military decision-making process. FM 3-05.20 and Graphic Training Aid (GTA) 31-01-003, *Detachment Mission Planning Guide*, provide additional information on planning.

ARMY INTERNATIONAL ACTIVITIES PROGRAM

3-8. The Army International Activities Program (AIAP) is the program that implements the Security Cooperation Guidance (SCG) from DOD. It supports the DOD security cooperation goals and provides the Army goals and objectives for Army security cooperation activities. Army International Activities support the NSS, the NMS, the regional strategies and the theater security cooperation plan (TSCP) of the combatant commanders, as well as the defense initiatives in the areas not assigned to the regional commands.

3-9. The AIAP is the policy and guidance link between the DOD SCG and the combatant command TSCP regarding security cooperation. It provides the guidance link to the Army component of the combatant command from the Army with policy and additional command guidance. Through this guidance, the Army component of the combatant command defines its role within the combatant command to effect security cooperation within that region and theater. Additionally, the Army component of the combatant command is also receiving direction from the combatant command regarding policy on security cooperation within that combatant command. The AIAP includes but is not limited to exchange programs, training programs, exercises, military-to-military contacts, and SA.

DEPARTMENT OF STATE GUIDANCE AND PLANNING

3-10. Generally, the DOS is the lead government agency for executing FID programs. Under Title 22 of the USC, DOS and DOD are responsible for SA to foreign countries. The DOS provides general program guidance, determines participating countries, approves specific projects, and integrates the military SA programs with other activities. Requirements for SA are resourced primarily by the HN and U.S. grants provided to DOD by executive transfers. DOD executes the SA program, identifies and prioritizes requirements, procures and delivers military equipment, and provides services. Within DOD, the DSCA provides overall direction, implementation, and supervision of approved SA and defense sales.

3-11. Policy, planning, and implementation of SA programs are incorporated into theater security cooperation planning, which includes planning for military FID operations. However, due to the different aspects of congressional oversight and funding of SA, DOS determines SA, and DOD implements it. SA policy flows from the President and eventually converges at the SAO. Generally, requirements for SA originate at the SAO in consultation with the HN and the GCC. The DOS puts forth policy to the Embassies and DOD. Throughout the policy flow, agencies produce plans that support SA policy and additional guidance issued throughout the process (for example, mission performance plans, TSCPs/strategy, and training plans) (Figure 3-1).

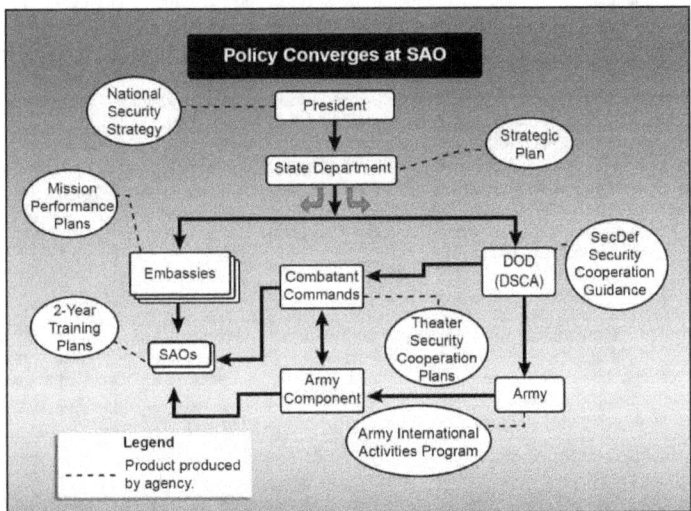

Figure 3-1. Army SA policy flow

THEATER PLANNING

3-12. Planning within the theater is the point at which DOS and DOD programs merge in the planning process to develop a program that fits the needs of the theater and its particular countries. The planning within the combatant commands is not completely uniform. Planners base their plans on higher-level guidance, priorities within the combatant command, and the resources available. It is a process that takes advantage of formal and informal arrangements with DOD, DOS, and interagency. The planning methodologies, assessments, and products developed may vary. However, a theater strategy and plan provides a basis for the activities to be conducted within that theater in support of that strategy. These activities include military FID operations in support of the theater strategy.

THEATER SECURITY COOPERATION PLANNING

3-13. The TSCP is primarily a strategic planning document intended to link GCC-planned regional engagement activities with national strategic objectives. Direction for the GCC is provided through the SecDef SCG and the JSCP. This guidance provides regional focus and security cooperation priorities. The SCG is implemented through the TSCP. The TSCP provides region-specific guidance, country guidance, and direction to further U.S. interests in the area of responsibility (AOR). Service component commanders and Commander, Special Operations Command (COMSOC), develop supporting security cooperation strategies to support the TSCP.

3-14. Combatant command planned and supported operations and activities produce multiple benefits in readiness, modernization, and security cooperation. However, peacetime military security cooperation activities must be prioritized to ensure efforts are focused on those that are of greatest importance, without sacrificing warfighting capability. The TSCP identifies the synchronization of these activities on a regional basis and illustrates the efficiencies gained from GCC security cooperation activities that support national strategic objectives. GCCs and executive agents will develop TSCPs for their assigned theaters or designated countries.

3-15. Within a combatant command, typically some type of planning conference, working group, or meeting is held annually. It is conducted to identify what type of SA, activities, and programs need to be implemented to support the SCG. Activities are prioritized based on the guidance from the annual meeting and are allocated to specific countries. Assessments can also be conducted on the previous year's activities to ensure validity, support to current guidance, and required updates.

3-16. The TSCP will specify all activities that will be conducted. Included within the TSCP are operational activities, combined exercises, combined training, SA, and HA. Planning, managing, and implementation of a security cooperation plan within the command are not identical. Each command may use various methods to develop a security cooperation plan. TSCP planning is a continuous process. The GCC TSCP strategic concept is normally updated biennially, and the activity annex is developed for the year of execution and the next seven years out. The TSCP planning process is a four-phase process (Figure 3-2, page 3-5). The phases are initiation, strategic concept development, activity annex development, and plan review. This process will occur in two stages.

Stage 1

3-17. In Phase 1, initiation, the GCCs receive planning guidance and planning tasks from the JSCP and the SecDef SCG. In Phase 2, strategic concept development, the GCC derives prioritized theater, regional, and country objectives. The strategic concept is developed. Resource requirements are identified to execute the strategy. The strategic concepts are reviewed and integrated and then collectively approved by the CJCS. The product is the completed strategic concept and is the completion of Stage 1.

Stage 2

3-18. Stage 2 begins with Phase 3, which is activity annex development. In this phase, security cooperation activities are identified. This phase describes in detail the activities to be conducted, to include operations, SA, exercises, and HA. Activities from this annex will be tasked as FID operations. Forces and resources are identified, the requirements are analyzed, and the shortfalls are identified. As required, the functional GCCs, Services, and other Defense agencies prepare and submit supporting and coordinating plans. The completed product is a TSCP. In Phase 4, plan review, the Joint Staff, Services, supporting GCCs, and the Office of the Under Secretary of Defense (Policy) (OUSD[P]) review the TSCPs. The TSCPs are integrated into the Global Family of Plans approved by the CJCS. The Global Family of Plans are then forwarded to the USD(P).

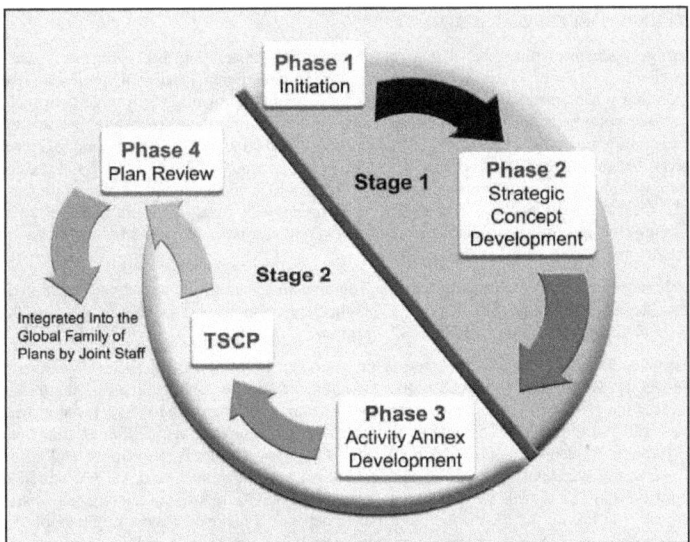

Figure 3-2. Theater security cooperation planning

THEATER SPECIAL OPERATIONS COMMAND AND JOINT SPECIAL OPERATIONS TASK FORCE

3-19. FID operations are predominately planned within the TSCP. SF takes a supporting role during the implementation of FID planning and operations within a theater. Theater special operations command (TSOC) representatives advise the GCC on the capabilities of SOF, provide SOF for employment, and integrate SOF fully into theater plans. TSOC representatives support the GCC by developing strategies to support the TSCP. This is done through planning, coordination, and recommendations that are included in the TSCP activity annexes.

3-20. On larger operations involving a joint special operations task force (JSOTF), SF may support a plan implemented by conventional military forces within a country to accomplish the combined U.S. and HN goals. SF units are required to conduct various missions in support of the FID program. The JSOTF is tasked to plan and conduct HN training. HN training can range from teaching advanced skills to training a force to conduct personal security detachment missions. The higher echelon tasks the JSOTF to conduct specific training requirements or maybe an end-state requirement that the JSOTF must plan and resource independently. Missions will vary in size and scope based on the combined U.S. and HN goals and the supporting role of SF units. Once JSOTF-level plans are developed, the special operations task force (SOTF) will develop training plans to support the FID program within their assigned area of operations (AO).

> *Note.* FM 3-05.20, *(C) Special Forces Operations (U)*, and GTA 31-01-003, *Detachment Mission Planning Guide*, include more information on planning.

3-21. When an operational detachment is tasked to conduct a FID mission, the detachment will plan that mission based on the military decision-making process. The following paragraphs will aid the detachment in planning and conducting a FID mission.

FOREIGN INTERNAL DEFENSE ASSESSMENT

3-22. Primarily, planners within the theater responsible for conducting FID programs assess what programs to conduct. DOS representatives work with foreign governments and DOD representatives work with foreign military personnel to develop programs that are consistent with U.S. foreign policy objectives and useful to the country concerned. The representatives developing the FID programs use the theater security cooperation planning process to assess currently implemented programs and exercises. The representatives developing the FID programs assess the previous programs for relevancy and success to the overall goals within the region. Assessments identify the effectiveness and strategic impact of the programs. To meet the goals of the U.S. security concerns and HN goals, the representatives review SA, exercises, training programs, and operational activities. They assess these programs on the basis of key trends, shortfalls, future opportunities, and challenges.

3-23. Specific personnel or forces are allocated to programs approved for implementation within a region. Exercises can be planned through the CJCS and GCC- or Service-sponsored training programs. The DSCA supports the implementation of approved U.S. SA programs.

3-24. Assessments to conduct a given mission or program can be completed at all levels of planning. At the tactical level, an SF unit can conduct a training assessment prior to conducting a training mission. The SF unit assesses the training requirements, personnel manning shortages, individual training needs, and equipment shortfalls of the HN unit. The unit will implement procedures to vet HN forces/units before they can receive training. (Appendix B provides a checklist for mission handoff procedures, and Appendix C provides a debriefing checklist for mission handoff.) Also, any personnel who are not vetted must be removed from training. DOS will vet personnel through the TSOC before conducting the mission. The primary purpose is to ensure the identification of personnel with a history of human rights violations. The U.S. policy is to prevent U.S. cooperation with governments of any country that engage in a consistent pattern of gross violations of internationally recognized human rights. Ideally, the site survey team gathers all this information. To properly conduct the training, the SF unit needs to determine or identify—

- The HN unit mission and mission-essential task list (METL), and its capability to execute them.
- The organizational tables for authorized personnel and equipment, and for personnel and equipment actually on hand.
- Any past or present foreign influence on training and combat operations using mobile training teams (MTTs), advisors, or available military equipment.
- The unit ability to retain and support acquired skills or training from past MTTs or foreign training missions.
- The organization and find out which leadership level is responsible for training the individual soldier. Does the HN have institutional training established and is it effective?
- Operational deficiencies during recent combat operations or participation in combined or joint exercises with the U.S. personnel.
- Maintenance status, to include maintenance training programs.
- The language or languages in which instruction will be conducted.
- The religious, tribal, or other affiliations within the HN forces that need to be considered.
- The potential security concerns with employing U.S. members in the HN training areas.

3-25. The SF unit also needs to review the relationship between the unit and the local population. It must determine if the unit is able to satisfy its administrative and logistics requirements without a negative impact on the civilian populace.

TRAINING PLAN

3-26. A key component of developing the training plan will be an agreement between the HN and the unit conducting the training. Training plans at the operational level will vary based on HN needs and unit training capabilities. An assessment for the training to be conducted can begin during the predeployment site survey (PDSS) (Appendix D). The considerations in the following paragraphs will aid the SF unit conducting unit-level training.

3-27. After completing the training assessment, the SF unit analyzes the prepared training plan and determines if changes are necessary. The SF unit develops FID tasks, conditions, and standards to train the HN forces. SF units tasked to train HN forces use the appropriate U.S. doctrine to attain the training goals. For example, they use battle drills and ARTEP MTPs, when applicable, to support HN training. HN training strategies must include multiechelon training whenever it is applicable. Multiechelon techniques save time and achieve synchronized execution of mission-essential tasks throughout the HN force. SF units assess the factors listed below when planning training programs and field exercises:

- HN's current level of training to determine if the training plan requires changes due to their level of proficiency or needs.
- Training facilities and areas based on projected training (for example, ranges and military operations in urban terrain sites).
- Proficiency of individuals and units in tactical operations and other skills required in IDAD operations involving intelligence, civil-military operations (CMO), and populace and resources control (PRC). Because of varied missions and limited resources, individuals and units require cross-training.
- Available equipment (for example, radios, weapons, and vehicles).
- C2 systems and logistics procedures, to include medical treatment and evacuation that stress decentralized operations over large areas.
- Cooperation with U.S. and HN intelligence agencies during operations and training exercises.
- Military civic action (MCA), particularly surveying needs and planning. Unit resources need a realistic assessment with the unit's primary mission in mind.
- Use of supporting CA and PSYOP units and the conduct of PSYOP and CA operations.
- Use of the unit to assist in PRC operations.
- Orientation on the terrain, climate, and unusual health requirements.

When developing a training plan, the SF unit must consider the training discussed in the following chapter.

This page intentionally left blank.

Chapter 4

Employment

In a FID mission, SF organize, train, advise, assist, and improve the tactical and technical proficiency of the HN forces. The goal is to allow HN forces to help themselves. The military presence in an HN's FID program could be at three levels—indirect, direct, or combat operations.

ROLE OF SPECIAL FORCES IN FOREIGN INTERNAL DEFENSE

4-1. U.S. military involvement in FID has traditionally focused on support of HN COIN efforts of allies and friendly nations. COIN remains an important aspect of military FID operations. However, the primary SF mission in FID is to organize, train, advise, assist, and improve the tactical and technical proficiency of the HN forces. The major difference in the way that SF and conventional forces conduct FID operations is in the area of advisory operations. Although conventional forces conduct a great deal of training in support of HN forces, they lack the capability to conduct effective advisory operations. As a force multiplier, SF units have and maintain advanced skills and capabilities (such as language) that enable them to conduct advisory operations with the HN for extended periods. Improved proficiency enables the HN forces to defeat internal threats to their stability, thereby limiting direct U.S. involvement. The emphasis is on training HN cadres, who will in turn train their compatriots. The capabilities that SF employ to perform their FID mission are those inherent to its UW mission. Only the operational environment is changed. United States Special Operations Command (USSOCOM) is the only combatant command with a legislatively mandated FID mission.

4-2. All SF personnel must understand the operational environment, to include political and legal implications of their operations. Legal considerations in planning and implementing FID programs are complex and subject to changing U.S. legislation. Commanders must keep their legal advisors involved in the planning process. Appendix E summarizes key legal aspects of FID activities.

4-3. An SF FID mission may require assets ranging from a single SF team to a reinforced SF group. In the early stages of a nation's need for assistance, the level of SF participation may be as small as one SFODA. In the more advanced stages, an SF company or battalion may establish an operational base (within or outside of country) and exercise OPCON of SF units. Operational and support elements may be assigned to the base on a rotational or permanent basis. When the entire SF group deploys to the country, it normally establishes a SOTF. The SOTF may then elect to establish one or more SF advanced operational bases (AOBs). SF units participate in a variety of operations to accomplish their FID mission. The HN needs and the U.S./HN agreements will dictate the quantity and level of support required to support the HN IDAD program.

4-4. There are various programs and exercises SF personnel can be involved with when supporting military FID operations. These missions can involve training, advising, and involvement in exercises sponsored through DOS and DOD initiatives.

TRAINING AND ADVISORY ASSISTANCE

4-5. SF elements may develop, establish, and operate centralized training programs for the supported HN force. SF can also conduct individual, leader, and collective training programs for specific HN units.

Subjects range from basic combat training and leader development to specialized collective training. SF can provide advisory assistance in two ways:

- SF teams may give operational advice and assistance to HN military or paramilitary organizations.
- Individual SF Soldiers may be assigned or attached to the SAO to perform advisory assistance duties on a temporary or permanent basis.

Note. In either case, assistance may be provided under the OPCON of the SAO chief in his role as the in-country U.S. defense representative or the TSOC, depending on the C2 arrangement.

TRAINING ASSISTANCE

4-6. The agreement negotiated between U.S. and HN officials provides the framework for the who, what, when, where, how, and why of military training assistance. Often, U.S. Army doctrine, as prescribed in applicable Army FMs, must be modified to fit the unique requirements of the HN forces being trained. Procedures may vary, but the fundamental techniques and thought process still apply.

4-7. In general, those skills, concepts, and procedures for FID taught to U.S. forces are also applicable to HN forces for IDAD. Training emphasis varies according to the HN requirements, force composition, and U.S./HN agreements. The training to be conducted depends on the situation and varies considerably. Existing military personnel, new military personnel, and/or paramilitary forces may receive training assistance.

4-8. HN counterpart personnel must be present with U.S. trainers. These counterparts will eventually conduct all the instruction and training without guidance from U.S. personnel. Initially, U.S. personnel may present all or most of the instruction with HN assistance, to include interpreters, if necessary. The goal of U.S. training assistance is to train HN personnel to conduct the training. U.S. trainers use the "train the trainer" concept. Figure 4-1 shows the general objectives of training programs under SA.

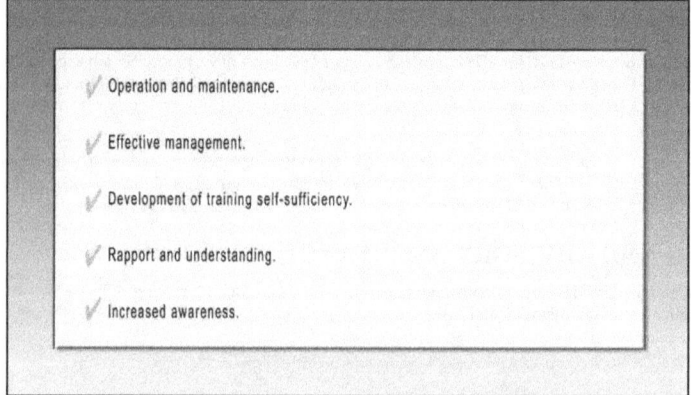

Figure 4-1. General objectives of training programs under SA

ADVISORY ASSISTANCE

4-9. Within DOD, the principal element charged with providing advisory assistance is the SAO. SF personnel may provide assistance in two ways: as an SF unit providing advice and assistance to an HN military or paramilitary organization or as an individual SF Soldier assigned or attached to the SAO. In either case, SF may be under OPCON of the SAO chief in his role as the in-country U.S. defense representative. However, SF will usually be under OPCON of a TSOC. The SAO includes all DOD elements, regardless of actual title, assigned in foreign countries to manage SA programs administered by DOD. The U.S. advisor may often work and coordinate with civilians of other U.S. Country Team agencies. When he does, he must know their functions, responsibilities, and capabilities since many activities cross jurisdictional borders. (Appendix F provides techniques for advisors.) The Country Team is composed of U.S. senior representatives of all USG agencies assigned to a country (Figure 2-3, page 2-6). Together, the SF advisor and his counterpart must resolve problems by means appropriate to the HN, without violating U.S. laws and policies in the process. SF advisors operate under very specific rules of engagement (ROE) with the purpose of ensuring that advisors remain advisors.

4-10. The SF advisor must understand the scope of SAO activities. He also must know the functions, responsibilities, and capabilities of other U.S. agencies in the HN. Because many SF activities cross the jurisdictional boundaries or responsibilities of other Country Team members, the SF advisor seeks other Country Team members to coordinate his portion of the overall FID effort.

4-11. Although refusing U.S. advisors, HN military leaders may request and receive other types of assistance such as air or fire support. To coordinate this support and ensure its proper use, U.S. liaison teams accompany HN ground maneuver units receiving direct U.S. combat support. Language-qualified and area-oriented SF teams are especially suited for this mission.

4-12. Figure 4-2 shows a typical structure for an SFODB. Figure 4-3, page 4-4, shows a possible C2 and advisory assistance relationship for a single SFODB deployed to provide advisory assistance to an HN brigade-sized unit. In Figure 4-3, the SFODB provides C2 systems, logistics for its subordinate SFODAs, and advisory assistance to the brigade-level echelon.

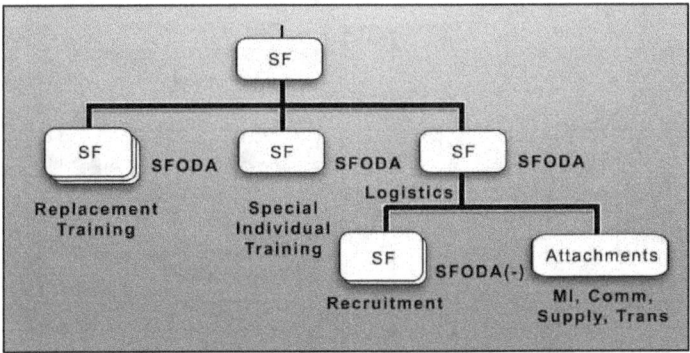

Figure 4-2. SFODB task organization for advisory assistance

4-13. Figure 4-4, page 4-4, shows another possibility for a C2 and advisory assistance relationship for a single SFODB deployed to provide advisory assistance to several individual HN battalion-sized units. In Figure 4-4, the SFODB only provides C2 systems and logistics for its subordinate SFODA. It does not have advisory assistance assigned for it.

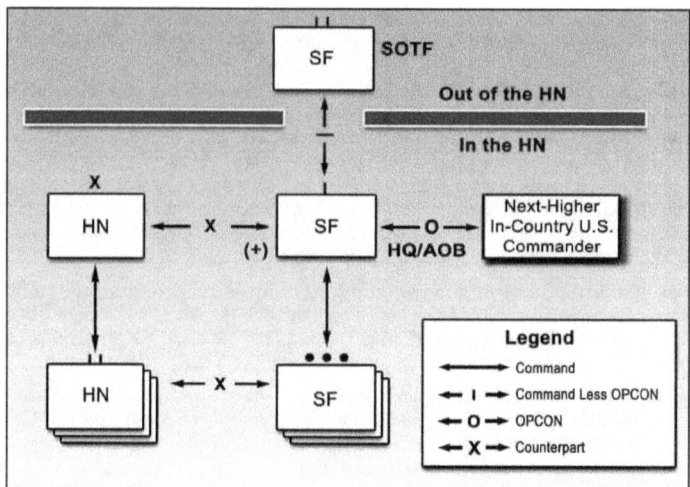

Figure 4-3. SFODB providing C2 systems, logistics, and advisory assistance

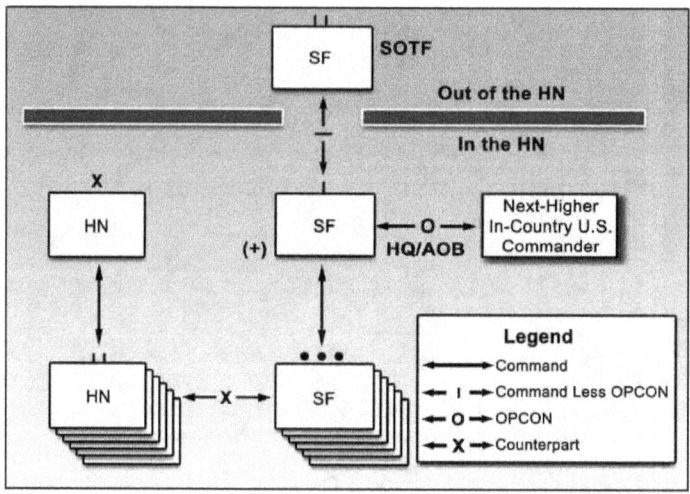

Figure 4-4. SFODB providing C2 systems and logistics for deployed SFODAs

4-14. Figure 4-5, page 4-5, shows a possibility for C2 and advisory assistance relationships for a single SF battalion deployed to provide advisory assistance to HN forces. In Figure 4-5, the SF companies are each responsible for providing advisory assistance to an HN brigade-sized unit.

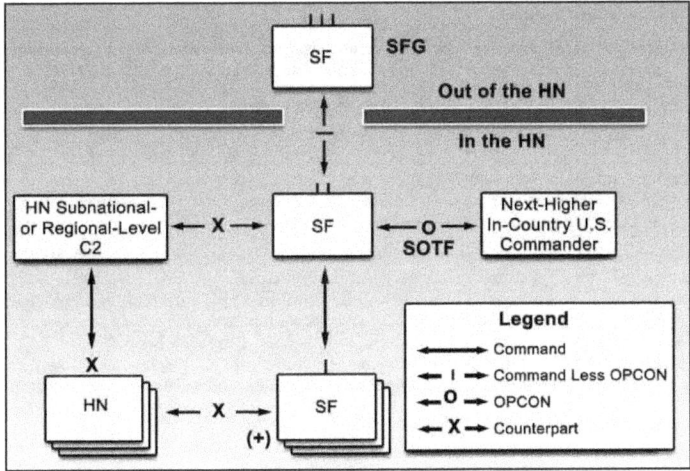

Figure 4-5. SFODB providing advisory assistance

SUPPORT FROM THE UNITED STATES FOR MILITARY FOREIGN INTERNAL DEFENSE OPERATIONS

4-15. Within an HN's FID program, the military instrument falls into three categories—indirect support, direct support, and combat operations. The levels are not constrained to a specific level of involvement. All levels of support can occur independently or simultaneously, and a specific level of escalation is not required. The type of support is based on an HN and a USG agreement. A successful FID program will consist of many of the elements listed in this chapter synchronized to fit the situation of a particular country.

INDIRECT SUPPORT

4-16. Indirect support builds strong national infrastructures through economic and military capabilities that contribute to self-sufficiency. These can include unit exchange programs, PEPs, individual exchange programs, and combination programs. Indirect support is provided to enhance an HN's capability to conduct its own operations.

Security Assistance

4-17. SA is a group of programs authorized by the Foreign Assistance Act of 1961, as amended, and the Arms Export Control Act (AECA) of 1976, as amended, or other related statutes by which the United States provides defense articles, military training, and other defense-related services by grant, loan, credit, or cash sales in furtherance of national policies and objectives. SA is a principal military instrument of the USG in assisting a friendly country along with other programs to assist a country with internal threats. The chief agencies involved in U.S. SA activities are DOS, Arms Transfer Management Group, DOD, JCS, GCC of the unified commands, SAO, and U.S. diplomatic missions. The following lists activities associated with indirect support to FID. SF personnel may be required to support many of these programs and exercises.

Foreign Military Financing Program

4-18. The principal means of ensuring America's security is through the deterrence of potential aggressors who would threaten the United States or its allies. Foreign Military Financing (FMF), the USG program for financing through grants or loans to acquire U.S. military articles, services, and training, supports U.S. regional stability goals and enables friends and allies to improve their defense capabilities. Foreign military sales (FMS) are made available under the authority of the AECA. Congress appropriates FMS funds in the International Affairs Budget, the DOS allocates the funds for eligible friends and allies, and the DOD executes the program. FMS help countries meet their legitimate defense needs, promote U.S. national security interests by strengthening coalitions with friends and allies, cement cooperative bilateral military relationships, and enhance interoperability with U.S. forces.

4-19. The Administration annually makes specific requests to Congress for the SA budget. The annual request is published in the Congressional Budget Justification (CBJ). The CBJ, prepared by the DOS, in coordination with the DSCA and other U.S. agencies, is presented to the Congress for those countries for which U.S. assistance is proposed. The Congress reviews the Administration's request and appropriates the funds for various international assistance programs; for example, Economic Support Fund (ESF), FMF, defense administration costs, voluntary peacekeeping operations (PKO), international military education and training (IMET), and Nonproliferation, Antiterrorism, Demining, and Related (NADR) programs.

Foreign Military Sales

4-20. The FMS program is the government-to-government method for selling U.S. defense equipment, services, and training. Responsible arms sales further national security and foreign policy objectives by strengthening bilateral defense relations, supporting coalition building, and enhancing interoperability between U.S. forces and militaries of friends and allies. These sales also contribute to American prosperity by improving the U.S. balance of trade position, sustaining highly skilled jobs in the defense industrial base, and extending production lines and lowering unit costs for key weapon systems. This program also fosters training opportunities for U.S. forces. For instance, Exercise IRIS GOLD is conducted with the Kuwaiti Army and managed under SA because it was an aspect of the U.S. program of FMS with Kuwait that incorporated numerous SF Soldiers, primarily from the 5th Special Forces Group (Airborne).

International Military Education and Training

4-21. The IMET program is an instrument of U.S. national security and foreign policy, and a key component of the U.S. SA program. The IMET program provides training and education on a grant basis to students from allied and friendly nations. In addition to improving defense capabilities, IMET facilitates the development of important professional and personal relationships. These relationships have proven to provide U.S. access and influence in a critical sector of society that often plays a pivotal role in supporting or transitioning to democratic governments. IMET's traditional purpose of promoting more professional militaries around the world through training has taken on greater importance as an effective means to strengthen military alliances and the international coalition against terrorism.

Counterterrorism Assistance

4-22. One program designed to assist nations in CT is the Counterterrorism Fellowship Program (CTFP). It is designed to assist regional commanders with their CT programs by funding foreign military officers and selected civilians to attend U.S. military educational institutions, outside the continental United States (OCONUS) mobile education and MTT courses, and selected regional centers for nonlethal training or other training and education permitted by Presidential and Congressional authorities. CTFP is designed to educate foreign military officers and selected civilian officials directly involved in the War on Terrorism to build CT capabilities and to provide friendly nations with the tools to enable them to sustain and grow their internal CT capabilities.

Humanitarian Mine Action Program

4-23. The Humanitarian Mine Action Program assists countries that are experiencing the adverse affects of uncleared landmines and other explosive remnants of war. Modern U.S. humanitarian mine action (HMA) began in 1986, when U.S. Army SF teams in southern Honduras trained Honduran Army engineers to clear landmines from agricultural land north of the Nicaraguan border. The program is directly managed by the GCCs and contributes to unit and individual readiness by providing unique in-country training opportunities that cannot be duplicated in the United States. A DOD component of the program is training indigenous personnel on mine clearing procedures, a train-the-trainer program. Over 4,000 indigenous personnel have benefited from this program. Training teams can include SF units, PSYOP, CA, explosive ordnance disposal, and conventional force engineers.

Joint and Multinational Exercises

4-24. Exercises conducted are designed to support the GCC's objectives within a specific theater or region. They are conducted to improve relations, enforce U.S. commitment to the region, improve interoperability with HN forces, and enhance U.S. warfighting skills. These exercises can be CJCS-, GCC-, and Service-sponsored events.

4-25. A program specific to SOF is the joint combined exchange training (JCET) program. The program is designed to train the SOF of the combatant command and is authorized under Section 2011, Title 10, United States Code (10 USC 2011), *Special Operations Forces: Training With Friendly Foreign Forces.* A historical mission of United States Special Forces (USSF) has been the training of foreign forces. USSF receive a bulk of their experience training HN forces through the JCET program. Added benefits to SF are regional familiarity, cross-cultural understanding, and access to numerous countries throughout the world not normally afforded to conventional forces.

Exchange Programs

4-26. Exchange programs primarily increase military contacts and increase military-to-military understanding and interoperability. Exchange programs can range from the exchange of a single person, such as the PEP, or an entire unit up to a battalion, such as the Reciprocal Unit Exchange Program with the United States. Operations INTRINSIC ACTION in Kuwait, BRIGHT STAR in Egypt, and EAGER LIGHT in Jordan maintained operating capabilities with Southwest Asian counterparts.

DIRECT SUPPORT

4-27. In direct support, U.S. forces provide direct assistance to the HN by actually conducting operations to support the civilian populace or the military. This support can be evaluation, training, limited information exchange, and equipment support. Direct support is usually funded by 10 USC authorities and may include training local military forces. The intent of direct support is to increase support to the HN, which may be in conjunction with indirect support. Direct support may not involve combat operations. The goal may be to keep U.S. forces from participating in combat operations, which may stem from political concerns, or to ensure the HN remains in the forefront of all operations to ensure or gain legitimacy. However, U.S. forces may become involved in combat operations when conducting direct support activities and will usually be guided by stricter ROE. The President of the United States must approve the conduct of combat activities by U.S. forces.

Civil-Military Operations

4-28. CMO are defined in JP 3-57, *Joint Doctrine for Civil-Military Operations.* This broad, generic definition denotes the decisive and timely use of military capabilities to perform traditionally nonmilitary activities. These activities include assisting host or friendly countries in bringing about political, economic, and social stability as they encourage the development of a country's materiel and human resources. FM 3-05.40, *Civil Affairs Operations,* further defines CMO as activities conducted by military units to enhance military effectiveness, support national objectives, and reduce the negative aspects of military operations on civilians. These activities include PRC, FHA, NA, support to civil administration (SCA), and civil information management (CIM).

4-29. CMO in FID support the internal development of the HN. They focus on the indigenous infrastructures and population in the operational area. Successful CMO will support the development of favorable attitudes, feelings, or behavior among the populace toward the HN IDAD projects.

4-30. SF Soldiers may become involved with CMO activities due to their association with civil and military leaders within their AOR through the conduct of their missions. SF can help CA units assist HN military forces develop effective CA programs that generate interest in the populace to support the IDAD programs of the HN government.

4-31. During mission analysis of a FID mission, the SF unit commander may determine that his team will require augmentation of a Civil Affairs team (CAT). Early CA augmentation will build on the SF unit understanding of the political, economic, social, religious, and cultural factors that will influence their operations in the HN. The CAT will be responsible for producing the SF unit CMO estimate and CA annex to the SF unit OPLAN. The CAT also assists the SF unit with a postdeployment area assessment to update area studies.

4-32. CA personnel working with the SF unit on a FID mission provide expertise and advisory capabilities in the area of CMO. They—

- Review U.S. SA program and HN IDAD goals.
- Plan CMO to support the HN plan.
- Plan CMO according to the three phases of insurgency described in this manual.
- Train HN military to plan, prepare for, and conduct MCA programs, PRC operations, NA, and other CMO appropriate to the IDAD of its country.
- Establish and maintain contact with nonmilitary agencies and local authorities.
- Identify specific CMO missions the HN military will conduct.
- Train on the tactics, techniques, and procedures (TTP) required to protect the HN from subversion, lawlessness, and insurgency.
- Develop indigenous individual leader and organizational skills to isolate insurgents from the civil population and protect the civil population.

4-33. CMO are the responsibility of military commanders at all levels. The successful military unit establishes a good working relationship with appropriate civil authorities and nonmilitary agencies in its AO. The SF unit must demonstrate how supported HN forces can integrate CMO into their military operations.

Foreign Nation Support

4-34. Foreign nation support (FNS) refers to the identification, coordination, and acquisition of HN or third-country resources to support military forces and operations. Although FNS is not a CA/CMO task, CA/CMO activities support it. HN or third-country resources include supplies, materiel, and labor that are not readily available to the military force by normal acquisition means. Purchase of these resources also adds to the local populace's trade and employment opportunities.

4-35. The SF unit helps the HN forces identify and acquire HN goods and services to support military operations. To accomplish this goal, the SF unit identifies projected shortfalls, determines what goods and services are available in the AO, and conducts negotiations for such support. Cultural awareness is extremely important in the negotiation process. Failure to follow locally accepted business principles could hurt efforts to establish rapport with the local populace and might play into the threat's propaganda campaign.

Foreign Humanitarian Assistance

4-36. FHA encompasses short-range programs such as disaster relief, noncombatant evacuation operations (NEOs), HCA, NA, and dislocated civilian (DC) operations aimed at ending or alleviating present human suffering. FHA is usually conducted in response to natural or man-made disasters, including combat. FHA is designed to supplement or complement the efforts of the HN civil authorities or agencies that have primary responsibility for providing FHA. The GCC's military strategy may include FHA to support FID as a component of the overall program to bolster the IDAD capability of the HN.

4-37. HCA programs can be very valuable to the GCC's support of FID programs while offering valuable training to U.S. forces. HCA programs are specific programs with funding authorized under Section 401,

Title 10, United States Code (10 USC 401), *Humanitarian and Civic Assistance Provided in Conjunction With Military Operations.* HCA programs assist the HN population in conjunction with a military exercise.

4-38. The SF unit, with its HN unit, may be directly involved in providing FHA to a needy populace. 10 USC 401 governs the use of U.S. military forces in HCA. Some forms of FHA may not extend to individuals or groups engaged in military or paramilitary activities.

4-39. The SF unit may also act as the coordinating or facilitating activity for FHA provided by the international nongovernmental organizations (NGOs) responding to the emergency needs of a community in the FID AO. The SF unit should get its HN military unit counterparts involved in this activity as early as possible to foster public support for the HN military.

Nation Assistance

4-40. NA is civil or military assistance (other than FHA) rendered to a nation by U.S. forces within that nation's territory during peacetime, crises or emergencies, or war based on agreements mutually concluded between the United States and that nation. NA operations support an HN by promoting sustainable development and growth of responsive institutions. The goal is to promote long-term regional stability. NA programs often include, but are not limited to, SA, FID, and 10 USC (DOD) programs, and activities performed on a reimbursable basis by federal agencies or international organizations. All NA activities are usually coordinated with the U.S. Ambassador through the Country Team. NA subtasks are SA, FID, and MCA.

4-41. MCA projects are designed to win support of the local population for government objectives and for the military forces in the area. MCA employs mostly indigenous military forces as labor and is planned as short-term projects. Projects must conform to the national plan and fit the development program for the area. Examples of these projects are farm-to-market roads, bridges, short-range education programs, basic hygiene, medical immunization programs, and simple irrigation projects.

4-42. For an MCA program to be successful, the local populace benefiting from the projects must have a voice in the selection of projects and the establishment of priorities. The SF unit must review (pretest) all projects with the populace before beginning the project. The SF unit must also conduct a posttest with the local people to determine whether the objectives were met. Failure to follow up can impact negatively on the overall IDAD mission in the area.

Civil Defense

4-43. Civil defense involves those measures taken to protect the populace and its property from harm should natural and man-made disasters occur. Civil defense is primarily the responsibility of government agencies. Civil-military problems are reduced when the government can control and care for its people. The effectiveness of civil defense plans and organization has a direct impact on other CMO.

4-44. SF unit support to civil defense could be as large as training and organizing a country's civil defense forces or simply training to enhance self-protection measures. MCA projects may assist the local populace in—

- Building new shelters or preparing existing facilities for emergency occupation.
- Planning and improving evacuation routes.
- Pursuing other measures that would save human life, prevent human suffering, or mitigate major destruction or damage to property.

Civil Information Management

4-45. Civil information is developed from data with relation to civil areas, structures, capabilities, organizations, people, and events within the civil component of the commander's battlespace that can be fused or processed to increase the situational awareness, situational understanding, or situational dominance of the DOD, interagency, international organizations, NGOs, and indigenous populations and institutions (IPI). CIM is the process whereby civil information is collected, entered into a central database, and internally fused with the supported element, higher headquarters (HQ), other USG and DOD agencies, international organizations, and NGOs to ensure the timely availability of information for analysis and the widest possible dissemination of the raw and analyzed civil information to military and nonmilitary partners throughout the AO. CIM subtasks are civil reconnaissance (CR) and civil information grid (CIG).

Other Considerations

4-46. Cultural characteristics in the AO are important to the local populace and require protection from military operations. The SF unit helps HN forces locate and identify religious buildings, shrines, and consecrated places, and recommends against using them for military purposes. The SF unit helps the HN forces determine methods and operational techniques that will be most acceptable to the populace and still allow for completion of the military mission.

4-47. If required, the SF unit, with its CA support, may support civil administration missions with the HN government. The SF unit helps HN military forces plan and conduct MCA. Since MCA is part of the overall U.S. SA program, formal agreements between the HN and the United States govern this support and CA activities.

4-48. PSYOP support to CMO primarily informs the populace about the many things the HN government and HN forces are doing for the people. Tactical loudspeaker teams, leaflets, and radio broadcasts are a few of the ways to let the people know about—

- What PRC measures are in effect.
- When certain PRC measures are no longer in effect.
- What civic action projects are being conducted in the area.
- What other programs are available for their benefit.

These PSYOP products can also keep the people abreast of the political, economic, and social situation in other parts of the country, and tactical and strategic successes of the government over insurgent forces.

4-49. The SF unit must observe the laws of armed conflict and ROE. The SF unit must quickly report human rights violations by HN or insurgent forces. The SF unit must be vigilant and act promptly, within its capability, to prevent or stop human rights violations. SF unit medical personnel may provide humanitarian treatment to civilians on an emergency-only basis, as their mission permits.

Psychological Operations

4-50. PSYOP must be an integral and vital part of an HN IDAD program. SF Soldiers may have to educate their HN counterparts in the value and role of PSYOP in FID. They must then advise and assist HN forces in developing and implementing an effective PSYOP program.

4-51. PSYOP can be used to gain the support of the people. Information activities target not only threat or foreign groups but also populations within the nation. Planners tailor PSYOP to meet specific needs for each area and operation. They evaluate the psychological impact of all military actions. Strict coordination and approval processes govern PSYOP programs. SF Soldiers must be aware that PSYOP are sensitive, strictly controlled activities that produce mid- to long-range results.

4-52. PSYOP support the achievement of U.S. national objectives and target specific groups. Examples of PSYOP goals for the main target groups in an insurgency follow. FM 3-05.30, *Psychological Operations*, provides more information on PSYOP.

4-53. PSYOP can support the mission by discrediting the insurgent forces to neutral groups, creating dissension among the insurgents themselves, and supporting defector programs. Divisive programs create dissension, disorganization, low morale, subversion, and defection within the insurgent forces. Also important are national programs to win insurgents over to the government side with offers of amnesty and rewards. Motives for surrendering can range from personal rivalries and bitterness to disillusionment and discouragement. Pressure from the security forces has persuasive power.

4-54. PSYOP should ultimately strive to identify the cause of insurgency behaviors or the contributing factors that are driving insurgency behaviors. By addressing the cause, PSYOP can target the perceptions and beliefs that are fueling the insurgency. PSYOP programs can also influence and change behaviors to indirectly deal with an insurgency such as the reporting of insurgent activity through various means.

Civilian Population

4-55. PSYOP support CMO activities by providing close and continuous information support. PSYOP maximize the return of CMO activities by passing instructions to the HN civilian populace that advertise the success or benefits of CMO programs to the populace. In the same vein, psychological actions within CMO programs reinforce the themes and messages of the PSYOP program by actively demonstrating the resolve of the HN and U.S. forces. PSYOP can also help establish HN command support of positive population control and protection from insurgent activities.

Host Nation Military/Paramilitary Forces

4-56. PSYOP can gain, preserve, or strengthen military support with emphasis on building and maintaining the morale and professionalism of military and paramilitary forces. The loyalty, discipline, and motivation of these forces are critical factors in FID.

Neutral Elements

4-57. PSYOP can support the FID mission by projecting a favorable image of the HN government and the United States. PSYOP can inform the international community of U.S. and HN intent and goodwill. PSYOP can also gain the support of uncommitted groups inside and outside the threatened nation by revealing the nature of the insurgency's subversive activities. PSYOP can bring international pressure to bear on any hostile power sponsoring the insurgency.

External Hostile Powers

4-58. PSYOP can convince the hostile power supporting the insurgents that the insurgency will fail. An effective PSYOP plan depends on timely information as well as intelligence and includes knowledge of the—

- History, culture, background, current environment, and attitudes of potential target groups.
- Insurgency's organization, motivation, and sources of conscription and material supply and how they are obtained.
- Strengths and weaknesses of ideological and political opponents.

4-59. The SF unit integrates the current PSYOP themes and objectives into its activities. The SF unit conducts itself (on and off duty) in a manner that has a positive, reinforcing psychological impact on the HN forces and the local populace.

Assessment

4-60. To determine PSYOP requirements during mission analysis, the SF unit assesses the psychological impact of its presence, activities, and operations in the AO. The SF unit reviews the OPLAN or operation order (OPORD) to ensure it supports U.S. and HN psychological objectives. This factor is critical. SF unit personnel analyze all official duties and consider the psychological impact on the populace when an SF unit participates in events such as military ceremonies, religious services, and social events. In addition, the SF unit must determine the practicality of planning and conducting training during national or religious holidays. The SF unit should consider requesting assets from the regional PSYOP battalion during predeployment and/or isolation to assist in mission analysis.

Protocol

4-61. SF unit members must conduct themselves in a proper, professional manner. They must observe local customs, local traditions, and U.S. Army standards of conduct. Each SF unit member must understand HN and local customs, courtesies, and taboos. As U.S. representatives to the HN, Americans can have a psychological impact on the mission by their actions (whether good or bad, or on- or off-duty). The supporting regional PSYOP battalion and the appropriate Country Team offices can assist the SF unit with cultural mores and development of a rapport-building program.

Image

4-62. Each SF unit operation integrates planned PSYOP activities to establish a favorable U.S. image in the HN and further the success of the SF unit mission. SF units coordinate with trained PSYOP assets to capitalize on positive mission successes. SF units can sometimes use HN and commercial media assets effectively to influence public opinion and pass information.

Assistance

4-63. The SF unit may have to advise and/or help HN forces in gaining or retaining the support of the local populace, discrediting the insurgents, and isolating the insurgents from the populace. The SF unit personnel influence the HN forces in conducting themselves in accordance with (IAW) acceptable military norms, mores, and professionalism. The SF unit trains the HN leadership in the advantages and techniques of maximizing public opinion in favor of the HN. The SF unit must support and assist, as much as possible, the HN mission to discredit the insurgents.

Support

4-64. The use of PSYOP assets and techniques will greatly enhance the effectiveness of CMO activities. The SF unit may advise and assist HN forces in how to use PSYOP to support their CMO objectives, and to integrate PSYOP capabilities into PRC measures.

Coordination

4-65. The SF unit must ensure the HN and U.S. mission approve local PSYOP activities and that the activities are consistent with U.S. national PSYOP goals and themes. Close coordination of military and CMO activities through HN agencies and the U.S. mission will ensure compliance with PSYOP guidance. Consistent monitoring of PSYOP activities in the AO will enhance the mission and ensure the commander's intent is met.

Military Training Support

4-66. The situation within an HN might require the need to train personnel beyond what is offered with indirect support. This training should remain focused on HN goals and the HN's IDAD strategy. However, the direct support training could focus on a particular aspect of the strategy with a focus on a particular issue.

Intelligence and Communication Sharing

4-67. Intelligence and communication sharing is extremely valuable in increasing an HN's capabilities. Levels of intelligence sharing must be carefully scrutinized. The sharing of intelligence and communications is sensitive. Disclosure of classified information must be authorized. Routine support can consist of training, limited information exchange, and equipment support. The degree of support must be balanced with the HN's capability to support that training and maintain the programs implemented. Support can range from counterintelligence (CI) elements to support of intelligence, surveillance, and reconnaissance (ISR) from the U.S. Air Force. Appendix G includes more information on intelligence operations.

Counterdrug Operations

4-68. U.S. law or DOD regulations impose many of the legal and/or regulatory constraints concerning HN CD operations. The SecDef may authorize support of federal, state, or local U.S. civilian law enforcement officials. This may include equipment support, advice, and training. Activities performed (including the provision of any equipment or facility, or the assignment or detail of any personnel) do not include or permit direct participation by a member of the Army, Navy, Air Force, or Marine Corps in a search, seizure, arrest, or other similar activity unless participation in such activity by such member is otherwise authorized by law (Section 375, Title 10, United States Code [10 USC 375], *Restriction on Direct Participation by Military Personnel*). The major HN-specific constraints are provided below:

- U.S. military forces are prohibited from accompanying HN forces on CD operations.
- Funds specially provided for HN support cannot be used for other security purposes.
- All operations must ensure the human rights of the citizens of the HN.
- Achieving CD objectives depends on the cooperation of the HN.

4-69. Assistance provided for HN CD efforts must be provided through SA and supported by CMO. Most of the CD efforts are supportive of U.S. FID initiatives.

4-70. HNs can obtain equipment from the United States to meet the internal threat to their security from lawlessness (drug trafficking). The training element of SA is a significant means of assistance for HNs. The GCC can provide training by SOF, conventional forces, or a combination of both. Following are the primary types of teams or programs that can be employed:

- *MTTs.* These forces are tailored for the training an HN requires.
- *Extended training service specialists (ETSSs).* These teams are employed over a longer period to help the HN train its own instructor cadre.
- *Deployment for training (DFT).* U.S. military units deploy to an HN for training to enhance their operational readiness and provide the added benefit of strengthening the operations of the HN.

COMBAT OPERATIONS

4-71. If the situation of the HN government deteriorates to the point that vital U.S. interests are in jeopardy, the President may commit U.S. forces in a combat role to effect a decisive change in the conflict. Direct U.S. military intervention can provide HN forces with the time and space to regain the strategic initiative and resume control of tactical operations. In this situation, the committed U.S. combat force is likely to find in-country SF teams with a myriad of formal and informal arrangements. The U.S. GCC fully exploits SF experience and contacts during the critical transition period when his forces are deploying into the country. He immediately exchanges liaison personnel with the proper SF HQ to exploit SF advice and assistance. The SF HQ provides all possible advice and assistance, to include—

- Situation and intelligence updates for incoming conventional force commanders and their staffs.
- Use of in-place SF elements for initial coordination with HN and U.S. mission agencies.
- Coalition support teams to facilitate integration of the HN forces into the overall plan.
- Real-time intelligence and operational reporting along with training status and operational capability assessment of HN units.
- Advisors to HN units to facilitate relief-in-place once specific objectives are met in selected sectors and/or AOs within the HN.
- Supervision of HCA efforts in remote areas to support the HN IDAD strategy.

4-72. SF may also be deployed to conduct FID operations within an HN with ongoing U.S. involvement in combat operations. This support can range from advising and training, to USSF conducting operations in support of HN combat operations to meet HN IDAD goals.

4-73. Generally, personnel participating in activities that fall under SA are restricted by law from combat. The AECA (Section 21) prohibits personnel providing defense services (including training) from performing duties of a combatant nature. Training and advising activities that may engage U.S. personnel in combat activities outside the United States are prohibited. Specifically, s (SATs) shall not engage in or provide assistance or advice to foreign forces in a combat situation. SATs are prohibited from performing operational duties of any kind except as may be required in the conduct of on-the-job training in the operation and maintenance of equipment, weapons, and supporting systems.

TERRORISM

4-74. If a nation is susceptible to or at a point of subversion, lawlessness, and insurgency, the possibility of terrorism or organizations intent on conducting terrorist activities could occur. U.S. military involvement in

FID has traditionally focused on COIN. Although much of the FID effort remains focused on this important area, U.S. FID programs may aim at other threats to an HN's internal stability, such as terrorism. Emphasis should be on helping the HN address the root cause of instability in a preventative manner rather than reacting to threats. Conditions such as unemployment, drug trafficking, violent crime, social unrest, and internal conflicts promote violent solutions like terrorism. Terrorism affects all aspects of a nation's defense and development. FID programs of all types, such as HA and especially CT, can prevent, reduce, or stop mitigating factors that can contribute to the beginning or spread of terrorism.

4-75. Fighting the War on Terrorism is an effort the United States cannot take on alone. It must be a global collaborative effort. DOS leads this collaborative effort and DOD supports it through numerous programs, which include military FID operations. These programs either directly or indirectly deter threats of terrorism within an HN and prevent the spread of a global threat, to include—

- Training and advising HN forces to deter crime and subversive activities.
- Intelligence- and communication-sharing to increase international awareness of terrorist organizations.
- CD support to stop or minimize narcoterrorism.

INFORMATION OPERATIONS

4-76. FM 3-13, *Information Operations: Doctrine, Tactics, Techniques, and Procedures*, defines IO as the "employment of the core capabilities of electronic warfare, computer network operations, psychological operations, military deception, and operations security, in concert with specified supporting and related capabilities, to affect or defend information and information systems, and to influence decision making." The purpose of IO is to affect the information environment to achieve information superiority over an adversary. Information superiority is the operational advantage derived from the ability to collect, process, and disseminate an uninterrupted flow of information while exploiting or denying an adversary's ability to do the same (FM 3-0, *Operations*). The ultimate targets of IO are the human decision-making processes and the attainment of information superiority, which enable U.S. forces to understand and act first. IO may involve complex legal and policy issues, requiring careful review and national-level coordination and approval. Additionally, IO require intelligence support for effective targeting and assessment. The IO cell on the joint force commander's (JFC's) staff deconflicts and synchronizes IO throughout the operations process to achieve unity of effort supporting the joint force. The IO cell is a critical element. Its presence ensures Army special operations forces (ARSOF) and joint SOF IO are integrated, coordinated, and deconflicted throughout the information environment. As appropriate, IO target or protect information, information-transfer links, information-gathering and information-processing nodes, and the human decision-making process through core, supporting, and related capabilities (Figure 4-6, page 4-15).

4-77. As in all military operations, IO are an integral part of the planning and execution of an HN's FID program. All organizations have information needs that must be met to operate effectively; the IO plan considers the capabilities and vulnerabilities that potential or existing insurgents have in the information environment. The IO plan seeks to gain an operational advantage over potential or existing insurgents by affecting their information content and flow at the right time and place in support of the overall FID program. IO, when properly synchronized and integrated, aid in legitimizing the FID program by helping to develop and maintain internal and international support while preempting potential or existing insurgent propaganda. An SF unit conducting operations in support of a FID program can expect a high degree of interaction with the HN military and civilian populace. This interaction creates psychological effects. The primary role of the SF unit in a FID program is to advise, train, and help HN forces protect its society from subversion, lawlessness, and insurgency. The following helps the SF unit accomplish its primary role in a FID program:

- Familiarity with the higher-level IO plan and with how their SF unit activities affect and support the plan.
- Knowledge of how the various IO capabilities can be synchronized to aid in positively shaping HN military and civilian populace awareness, perceptions, and understanding.

Note. FM 3-13 has additional information on IO.

CORE CAPABILITIES:

Operations Security

Psychological Operations

Military Deception

Electronic Warfare

Computer Network Operations

SUPPORTING CAPABILITIES:

Information Assurance

Physical Security

Physical Attack

Counterintelligence

Combat Camera

RELATED CAPABILITIES:

Civil-Military Operations

Public Affairs

Figure 4-6. IO capabilities

This page intentionally left blank.

Appendix A

Insurgency and Counterinsurgency

An insurgency is an organized movement aimed at the overthrow of a constituted government using subversion and armed conflict. In some cases, however, the goals of an insurgency may be more limited. For example, the insurgency may intend to break away a portion of the nation from government control and establish an autonomous state within traditional ethnic or religious territorial bounds. The insurgency may also intend to extract limited political concessions unattainable through less violent means. COIN is defined as those military, paramilitary, political, economic, psychological, and civic actions taken by a government to defeat insurgency (FM 1-02).

NATURE OF INSURGENCIES

A-1. Insurgencies generally follow a revolutionary doctrine and use armed forces as an instrument of policy. An insurgency is a protracted politico-military struggle designed to weaken government control and legitimacy while increasing insurgent control and legitimacy—the central issues in the insurgency. Each insurgency has its own unique characteristics based on its strategic objectives, its operational environment, and available resources. Insurgencies normally seek to overthrow the existing social order and reallocate power within the country.

PHASES OF INSURGENCIES

A-2. An insurgency may be classified into three general phases according to the level of intensity. Typically, successful insurgencies pass through common phases of development. Not all insurgencies experience every phase, and progression through all phases is certainly not a requirement for success. The same insurgent movement may be in another phase in other regions of a country. Successful insurgencies can also revert to an earlier phase when under pressure, resuming development when favorable conditions return.

A-3. Some insurgencies depend on proper timing for their success. Because of their limited support, their success depends on weakening the government's legitimacy so that it becomes ineffective. Then, an opportunity to seize power exists. When these insurgencies move to seize power, they expose their organization and intentions. If they move too early or too late, the government may discover their organization and destroy it. Timing is critical.

PHASE I

A-4. Phase I ranges from circumstances in which subversive activity is only a potential threat, latent or incipient, to situations in which subversive incidents and activities occur with frequency in an organized pattern. Phase I involves no major outbreak of violence or uncontrolled insurgency activity.

A-5. In those nations where a potential insurgency problem exists and where U.S. interests so dictate, an SA program may be designed. SA programs support the total U.S. effort to reduce the causes of insurgency. Initially, such a program will provide a continuing assessment of the threat and allow work toward strengthening the indigenous capacity to combat insurgency. U.S. military intelligence (MI) activity in this phase is primarily a CI effort involving the assessment of such potential hostile threats as terrorism, espionage, and sabotage to U.S. national security interests and the reliability of non-U.S. military resources.

A-6. If an SAO does not exist, the nation concerned should be encouraged to obtain appropriate assistance by requesting establishment of an SAO or requesting JCET programs. The theater commander's security cooperation strategy for the region defines the conduct of these programs. The mission should include U.S. personnel specially trained in military assistance. Personnel trained specifically for other FID activities may serve as augmentees for the mission. By these means, HN forces can have appropriate training to better facilitate their dealing with the problem.

PHASE II

A-7. Phase II is reached when the subversive movement, having gained sufficient local or external support, initiates organized guerrilla warfare or related forms of violence against the established authority. In situations where insurgency develops to more serious proportions, U.S. efforts may be expanded to include—

- Necessary equipment and training.
- Forces specifically trained for activities in FID.
- Instructor personnel.

Note. Under some circumstances, unit advisors may also be included.

PHASE III

A-8. The situation moves from Phase II to Phase III when the insurgency becomes primarily a war of movement between organized forces of the insurgents and those of the established authority. During a period of escalated insurgency, the United States may expand its assistance at the request of the host government. This assistance may include selected U.S. conventional forces. Nevertheless, the HN government will be expected to provide the bulk of the combat forces required in dealing with the situation. It is critical for the HN forces to remain at the forefront of the effort to ensure they remain legitimate in the eyes of the HN populace. This effort can be supported through the IO plan.

CAUSES OF DYNAMICS OF INSURGENCIES

A-9. Insurgencies may arise when the populace perceives that the government is unable or unwilling to redress their issues or the demands of important social groups. These groups band together and begin to use violence to change the government's position. Insurgencies are often a coalition of disparate forces united by their common enmity for the government. To be successful, an insurgency must develop unifying leadership, doctrine, organization, and strategy. Only the seeds of these elements exist when an insurgency begins. The insurgents must continually nurture and provide the necessary care if the insurgency is to mature and succeed.

A-10. Insurgencies succeed by mobilizing human and materiel resources to provide active and passive support for their programs, operations, and goals. Mobilization produces workers and fighters, raises funds, and acquires the necessary weapons, equipment, and supplies. Mobilization grows out of intense, popular dissatisfaction with existing political and social conditions. The active supporters of the insurgency consider these conditions intolerable. The insurgent leadership articulates its dissatisfaction, places the blame on government, and offers an alternative. The insurgent leadership then provides organizational and management skills to transform disaffected people into an effective force for political action. Ultimately, the insurgents need the active support of a majority of the politically active people and the passive acquiescence of the majority.

FRAMEWORK FOR ANALYSIS OF INSURGENCIES

A-11. There are eight dynamics common to most insurgencies:

- Leadership.
- Ideology.
- Objectives.
- Environment and geography.
- Internal support.

- External support.
- Phasing and timing.
- Organizational and operational patterns.

A-12. These dynamics provide a framework for analysis that can reveal the strengths and weaknesses of the insurgency. Although analysts can examine these elements separately, they must understand how they interact to fully understand the insurgency.

LEADERSHIP

A-13. Insurgency is not simply random political violence; it is directed and focused political violence. It requires leadership to provide vision, direction, guidance, coordination, and organizational cohesiveness. The leaders of the insurgency must make their cause known to the people and gain popular support. Their key tasks are to break the ties between the people and the government and to establish the credibility of their movement. They must replace the legitimacy of the government with that of their own. Their education, background, family, social connections, and experiences shape how they think and how they will fulfill their goals. These factors also help shape their approach to problem solving.

A-14. Leadership is a function of organization and personality. Some organizations de-emphasize individual personalities and provide for redundancy and replacement in decision making. These mechanisms produce collective power and do not depend on specific leaders or personalities to be effective. They are easier to penetrate but more resistant to change. Other organizations may depend on a charismatic personality to provide cohesion, motivation, and a rallying point for the movement. Organizations led in this way can produce decisions and initiate new actions rapidly but are vulnerable to disruptions if key personalities are removed or co-opted.

IDEOLOGY

A-15. To win, the insurgency must have a program that explains what is wrong with society and justifies its insurgent actions. It must promise great improvements after the government is overthrown or if its goals are met. The insurgency uses ideology to offer society a goal. The insurgents often express this goal in simple terms for ease of focus. Future plans of the insurgency must be vague enough for broad appeal and specific enough to address important issues.

A-16. The ideology of groups within the movement may indicate differing views of strategic objectives. Groups may have ideological conflicts that need to be resolved before an opponent can capitalize on them. Ideology may suggest probable objectives and tactics. It greatly influences the insurgent's perception of his environment. This perception of the environment in turn shapes the organizational and operational methods of the movement.

OBJECTIVES

A-17. Effective analysis of an insurgency requires interpretation of the objectives possibly pursued by the insurgents, to include—

- *Strategic objective.* The strategic objective is the insurgent's desired end state; that is, how the insurgent will use the power once he has it. The replacement of the government in power is only one step along this path. However, it will likely be the initial focus of efforts. Typically, the strategic objective is critical to cohesion among insurgent groups. It may be the only clearly defined goal of the movement.
- *Operational objective.* Operational objectives are those the insurgents pursue as part of the overall process of destroying government legitimacy and progressively establishing their desired end state.
- *Tactical objective.* Tactical objectives are the immediate aims of insurgent acts; for example, the dissemination of PSYOP products or the attack and seizure of a key facility. These actions accomplish tactical objectives that lead to operational goals. Tactical objectives can be psychological and physical in nature. For example, legitimacy is the center of gravity for the

insurgents and the counterinsurgents. Legitimacy is largely a product of perception. Consequently, it can be the principal consideration in the selection and attainment of tactical objectives.

ENVIRONMENT AND GEOGRAPHY

A-18. Environment and geography, including cultural and demographic factors, affect all participants in a conflict. The manner in which insurgents and counterinsurgents adapt to these realities creates advantages and disadvantages for each. The effects of the environment and geography are most visible at the tactical level where they are perhaps the predominant influence on decisions regarding force structure, doctrine, and TTP.

INTERNAL SUPPORT

A-19. The population's support is fundamental to the success of both the insurgency and COIN operations. The population's support for the insurgency, even its neutrality, will allow the insurgents the freedom of movement and the ability to rest, refit, and recruit. In the same manner, COIN operations require the popular support of the people to acquire the necessary information to plan, conduct, and continue its operations. These opposing forces are in a constant struggle to gain and maintain the support of the people and thereby create an internal support mechanism.

EXTERNAL SUPPORT

A-20. Historically, some insurgencies have done well without external support. However, recent examples, such as Vietnam, Nicaragua, Afghanistan, and Iraq show that external support can accelerate events and influence the outcome. External support can provide political, psychological, and material resources that might otherwise be limited or unavailable. Four types of external support are—

- *Moral*: Acknowledgement of the insurgent as just and admirable.
- *Political*: Active promotion of the insurgents' strategic goals in international forums.
- *Resources:* Money, weapons, food, advisors, and training.
- *Sanctuary*: Secure training plus operational and logistical bases.

A-21. Accepting external support can affect the legitimacy of insurgents and counterinsurgents. It implies the inability to sustain oneself. In addition, the country or group providing support attaches its legitimacy along with the insurgent or the counterinsurgent group it supports. The consequences can affect programs in the supporting nation wholly unrelated to the insurgent situation. However, adverse consequences can be alleviated through anonymous contributions that are channeled through various sources before reaching the insurgent group.

PHASING AND TIMING

A-22. Insurgencies often pass through common phases of development. The conceptualization generally followed by insurgents is drawn from that postulated by Mao Tse-tung. Regardless of its provenance, movements as diverse as Communist or Islamic insurgencies have used the Maoist conceptualization because it is logical and based upon the mass mobilization emphasis. It states that insurgents are first on the strategic defensive (Phase I), move to stalemate (Phase II), and finally go over to the offensive (Phase III). Strategic movement from one phase to another incorporates the operational and tactical activity typical of earlier phases. It does not end them.

ORGANIZATIONAL AND OPERATIONAL PATTERNS

A-23. Insurgencies develop organizational and operational patterns from the interaction of many factors. As a result, each insurgency organization is unique. However, knowing the commonly accepted general patterns or strategies of insurgency helps in predicting the tactics and techniques they may employ against the supported government.

INSURGENT STRATEGIES

A-24. There are three general strategies of insurgency: traditional, "foco" (Spanish word meaning focus or focal point), and mass-oriented. The following paragraphs discuss these strategies.

TRADITIONAL INSURGENCY

A-25. A traditional insurgency normally grows from very specific grievances and initially has limited aims. It springs from tribal, racial, religious, linguistic, or other similarly identifiable group. The insurgents perceive that the government has denied the rights and interests of their group and work to establish or restore them. They frequently seek withdrawal from government control through autonomy or semi-autonomy. They seldom specifically seek to overthrow the government or control the whole society. They generally respond in kind to government violence. Their use of violence can range from strikes and street demonstrations to terrorism and guerrilla warfare. These insurgencies may cease if the government accedes to the insurgents' demands. The concessions the insurgents demand, however, are so great that the government concedes its legitimacy along with them.

Huk Rebellion

The Huk rebellion in the Philippines can be considered a traditional insurgency despite its Communist origin. The Huks first surfaced as an armed force resisting the Japanese occupation during World War II. After the war, when other resistance bands disarmed, the Huks did not. After the American liberation, the Huks saw a chance to seize national power at a time when the newly proclaimed Philippine Republic was in obvious distress because of a monetary crisis, graft in high office, and mounting peasant unrest. By 1950, the Huks had built a force of 12,800 armed guerrillas with thousands of peasant supporters on central Luzon. They were defeated in a series of actions by the Armed Forces of the Philippines led by Ramon Magsaysay. By 1965, they were nearly extinct and down to 75 members. Largely agrarian, the Huks do not view the government as totally in need of replacement but that many of the people in it need replaced. Recently, the Huk movement has been gaining popular support on the island of Luzon.

FOCO INSURGENCY

A-26. A foco is a single, armed cell that emerges from hidden strongholds in an atmosphere of disintegrating legitimacy. In theory, this cell is the nucleus around which mass popular support rallies. The insurgents build new institutions and establish control on the basis of that support. For a foco insurgency to succeed, government legitimacy must be near total collapse. Timing is critical. The foco insurgency must mature at the same time the government loses legitimacy and before any alternative appears. The most famous foco insurgencies were those led by Fidel Castro and Che Guevara. The strategy was quite effective in Cuba because the Batista regime was corrupt and incompetent. The distinguishing characteristics of a foco insurgency are the—

- Deliberate avoidance of preparatory organizational work. The rationale is based on the premise that most peasants are intimidated by the authorities and will betray any group that cannot defend itself.
- Development of rural support, as demonstrated by the ability of the foco insurgency to strike against the authorities and survive.
- Absence of any emphasis on the protracted nature of the conflict.

Castro's Junta

In 1952, Fidel Castro began his revolutionary movement in Cuba. After an unsuccessful attack of Fort Moncada, he was imprisoned. Upon release in 1955, he fled to Mexico to train a new group of guerrilla warriors. In 1956, Castro and 82 of his followers returned to Cuba on a yacht. Of this group, only 12 of Castro's followers made their way to the Sierra Maestra mountain range. From his remote mountain base, Castro established a 100- to 150-man nucleus. As Castro's organization grew, small unit patrols began hit-and-run type operations. While Castro continued to expand his area of influence, the popularity of the corrupt Batista government waned. In May of 1958, the government launched an attack on the Sierra Maestra stronghold. Castro withdrew deeper into the mountains, while spreading his message on national reform. Batista's continuing repression of the country led to general strikes and continuing growth in popular support for Castro's small cell of revolutionaries. Finally, Batista fled the country on 1 January 1959, and Castro established a junta and became the Prime Minister and President.

MASS-ORIENTED INSURGENCY

A-27. A mass-oriented insurgency aims to achieve the political and armed mobilization of a large popular movement. Mass-oriented insurgencies emphasize creating a political and armed legitimacy outside the existing system. They challenge that system and then destroy or supplant it. These insurgents patiently build a large armed force of regular and irregular guerrillas. They also construct a base of active and passive political supporters. They plan a protracted campaign of increasing violence to destroy the government and its institutions from the outside. Their political leadership normally is distinct from their military leadership. Their movement normally establishes a parallel government that openly proclaims its own legitimacy. They have a well-developed ideology and decide on their objectives only after careful analysis. Highly organized, they mobilize forces for a direct military and political challenge to the government using propaganda and guerrilla action. The distinguishing characteristics of a mass-oriented insurgency are—

- Political control by the revolutionary organization, which ensures priority of political considerations.
- Reliance on organized popular support to provide recruits, funds, supplies, and intelligence.
- Primary areas of activity, especially in early phases, in the remote countryside where the population can be organized and base areas established with little interference from the authorities.
- Reliance upon guerrilla tactics to carry on the military side of the strategy. These tactics focus on the avoidance of battle, except at times and places of the insurgents' choosing, and the employment of stealth and secrecy, ambush, and surprise to overcome the initial imbalance of strength.
- A phased strategy consisting first of a primarily organizational phase in which the population is prepared for its vital role. In the second phase, "armed struggle" is launched and the guerrilla force gradually builds up in size and strength. The third phase consists of mobile, more conventional warfare. Conceptually, this third phase is accompanied by a popular uprising that helps overwhelm the regime. It is in concept a "protracted" war.

Advising El Salvador Military

For 12 years, beginning in 1979, the United States assisted the El Salvador military in becoming a more professional and effective fighting force against the Communist-backed Farabundo Marti National Liberation Front. A U.S. military group assisted the El Salvadoran army by establishing a facility for basic and advanced military training. SF advisors, primarily from the 7th Special Forces Group, served with El Salvadoran units to support small-unit training and logistics. The advisors helped the El Salvadoran military become more professional and better organized, while advising in the conduct of pacification and counterguerrilla operations. Advisors were also present at the brigade levels assisting in operations and intelligence activities. From 1985 to 1992, just over 140 SF officers and noncommissioned officers (NCOs) served as advisors to a 40-battalion army. From a poorly staffed and led force of 8,000 soldiers in 1980, SF trainers created a hard-hitting COIN force of 54,000 by 1986. U.S. forces supported U.S. interests by creating an effective COIN force that fought the guerrillas to a standstill and established the groundwork for a negotiated settlement by 1991.

INITIATING EVENT

A-28. It does not follow that an insurgency will erupt if the preconditions for an insurgency are satisfied. The conflict must await an initiating event. An initiating event mobilizes the energies of the discontented and directs them toward violent action. Its impact is more psychological than physical and need not follow immediately after the event. The event may have little significance, but an elite group or an organization may, at some time, give it special significance. Possible initiating events include—

- An event that gains symbolic significance. This event may be an economic or social disaster, a particularly antagonizing action by the regime, or a heroic act of defiance by an individual.
- An event that forces action, such as an invasion by a foreign power.
- The emergence of a charismatic leader (for example, Fidel Castro or Mao Tse-tung).
- The perception of a tactical or strategic advantage by revolutionary elite.
- The decision by revolutionary elite to issue a call to arms.
- The influence of foreign agents or propaganda.

A-29. If a situation is explosive, almost any event may serve as an initiating event. Its correct timing may also produce a flood of events in a short period, making it hard to point to a single event as the act that initiated the struggle. Thus, it may be more helpful to think of a series of acts as an initiating event. Initiating events may be historical, with the insurgents recalling the event for the populace. This technique frees the insurgent from waiting for a proper event to occur.

Note. FM 3-05.201, *Special Forces Unconventional Warfare Operations*, has additional information on insurgency.

TACTICAL COUNTERINSURGENCY OPERATIONS

A-30. Tactical COIN operations reduce the insurgent threat or activity in the area and provide a favorable environment for the HN IDAD program. These objectives are complementary. When the insurgent threat is reduced, internal development can begin. When it works, internal development alleviates the causes of dissatisfaction that gave rise to the insurgency by depriving the insurgent of popular support and a reason for fighting. Basic considerations for successful COIN operations are training, intelligence, a framework for combat, and a well-defined C2 arrangement by which the civilian government exercises control and coordination of all COIN operations.

A-31. There may be a need for tactical operations inside or near an urban area to defeat an insurgent attack. Any insurgent effort to seize and hold an urban area will probably involve operations in nearby areas as well. When the police or other internal security forces cannot cope with the attack inside the urban area, military forces can participate. These forces can set up security around the urban area and deny the insurgents reinforcements or support.

MILITARY FORCES

A-32. When military forces reinforce police units to defeat insurgent forces inside an urban area, they require close control and coordination. The military forces should make every attempt to empower HN forces to remain at the forefront of operations to build or maintain legitimacy in the eyes of the populace. At times they may only require military units to conduct outer cordons or act as a quick reaction force (QRF) for police operations. As soon as the police force can manage the situation, the military forces withdraw.

INSURGENT FORCES

A-33. When insurgent forces seize an urban area, proper authorities evaluate (from a tactical and psychological aspect) whether to recapture the area by using major military force or other techniques. The probable psychological impact on the enemy, noncombatant civilians, and friendly troops influences the amount of force and specific techniques used to recapture an area. The safety of civilians and friendly troops, probable damage to property, and the military forces available are considerations. The principle of minimum essential force will help reduce casualties in the noncombatant civilian population.

COORDINATION

A-34. Riot-control munitions and nonlethal weapons systems can be used against targets so that military forces can close with and capture the enemy with minimum injury to the noncombatants. As such, military operations must be coordinated with the civilian police.

MISSION IMPACT CONSIDERATIONS

A-35. Subordinate commanders have maximum flexibility in the execution of their missions but receive specific responsibilities and enough guidance to ensure a coordinated effort. Events may cause rapid changes to COIN OPLANs and allow units to use their resources against exposed guerrilla forces.

A-36. Maintaining high morale in units engaged in COIN operations presents problems different from those in limited and conventional operations. Operating against an elusive force that seldom offers a clear target and where tangible results are seldom obtained requires continuous troop indoctrination and training.

A-37. During independent, prolonged missions, unit support depends on the ingenuity, courage, and tenacity of commanders and staffs at all echelons. Command and staff action in COIN operations emphasizes—

- Detailed planning of small-scale, decentralized operations.
- Covering extended distances.
- Extensive contingency planning for the use of reserves and fire support.
- Deception operations.
- The use of electronic warfare (EW) operations.
- Detailed planning and coordination of activities with nonmilitary government officials.

A-38. In COIN operations, command and staff action also emphasizes detailed coordination and direction of the intelligence collection effort. These actions take place by—

- Coordinating with HN and U.S. intelligence agencies and HN regular and paramilitary forces.
- Using combat forces and EW intelligence elements, to include radar and remotely monitored sensors and other technical surveillance systems.

- Using local people in the development of intelligence collection systems.
- Systematically and thoroughly interrogating prisoners and suspects.

A-39. In addition, command and staff action in COIN operations emphasizes incorporating and monitoring government internal development programs in the OPLAN. These actions include—

- Preparing and executing integrated plans that include IO, CA, PRC, and PSYOP.
- Operating with and assisting HN military, paramilitary, and police forces.
- Integrating logistics functions, especially aerial resupply, into all planning.

MILITARY OPERATIONS IN COUNTERINSURGENCY

A-40. The objective of military operations in COIN is to provide a secure environment in which balanced development can occur. Military operations should not be independent military actions aimed solely at destroying insurgent combat forces and their base areas. Military operations must be part of a synchronized effort to gain broader goals. The SF team commander must convince his counterpart to integrate intelligence, IO, CA, and PSYOP activities into every military operation. SF advisors and their HN counterparts must be aware of the impact their actions have on the populace and other IDAD programs. SF personnel have extensive knowledge of UW, language, and culture that makes them uniquely qualified to advise and assist the HN in how to organize, equip, train, sustain, and employ combat forces in COIN operations. SF may participate in the types of operations described below.

CONSOLIDATION OPERATIONS

A-41. Consolidation operations are long-term population security operations conducted in territory generally under HN government control. Their purpose is to—

- Isolate the insurgents from the populace.
- Protect the populace from insurgent influence.
- Neutralize the effects of the insurgents on the population.
- Neutralize the insurgent infrastructure.

A-42. The people are unlikely to support the HN government fully until the government provides enough long-term security to free its people from the fear of insurgent reprisals. Consolidation operations accomplish these objectives.

STRIKE OPERATIONS

A-43. Strike operations are short-duration tactical operations conducted in contested or insurgent-controlled areas (unlike consolidation operations). Strike operations are primarily offensive operations. Small, highly mobile combat forces operate in dispersed formations to locate and fix the insurgent force. Upon locating the insurgent force, strike force commanders have their forces attack, pursue, and destroy it. If contact is lost, the strike forces resume aggressive patrolling to reestablish contact and destroy the insurgent force before it can rest, reorganize, and resume combat operations. The purpose of strike operations is to destroy insurgent forces and base areas, isolate insurgent forces from their support, and interdict insurgent infiltration routes and lines of communications (LOCs).

REMOTE AREA OPERATIONS

A-44. Remote area operations take place in insurgent-controlled or contested areas to establish islands of popular support for the HN government and deny support to the insurgents. They differ from consolidation operations in that they do not establish permanent HN government control over the area. Ethnic, religious, or other isolated minority groups may populate remote areas. They may be in the interior of the HN or near border areas where major infiltration routes exist. Remote area operations normally involve specially trained paramilitary or irregular forces. SF teams support remote area operations to interdict insurgent activity, destroy insurgent base areas, and demonstrate that the HN government has not conceded control to the insurgents. They also collect and report information on insurgent intentions in more populated areas.

PSYOP and CA programs help in obtaining local support for remote area operations. Success is more likely if—

- A significant segment of the local population supports the program.
- The HN recruits local personnel for its remote area paramilitary or irregular force.
- HN forces conduct remote area operations to interdict infiltration routes in areas nearly devoid of any people. In this case, SF teams advise and assist irregular HN forces operating in a manner similar to that of insurgents but with access to superior logistics resources.

BORDER OPERATIONS

A-45. HN police, customs, or paramilitary border forces should be responsible for border security. However, the threat may require combat-type border operations, particularly in remote areas. SF teams advise and assist HN forces assigned to prevent or interdict the infiltration of insurgent personnel and materiel across international boundaries. The intent is to isolate insurgent forces from their external support, to include external sanctuaries. Secondary purposes are to locate and interdict insurgent land infiltration routes, destroy insurgent forces and base areas in areas adjacent to the border, and collect and report information on insurgent capabilities and intentions. Border operations normally require restrictive PRC measures. These PRC measures are particularly annoying to border tribal and ethnic groups that do not recognize the international boundary. The HN government must make a continuing PSYOP effort to gain and maintain the loyalty of the affected populace.

URBAN AREA OPERATIONS

A-46. Clandestine insurgent activity may be extensive in urban areas. Activities may include terrorism, sabotage, PSYOP, and political, organizational, intelligence, and logistic operations. This insurgent activity may strain the capabilities of police and other civil authorities. Police, internal security, and other HN government organizations will be high-priority targets for the insurgents. The insurgents normally try to exploit local civilian organizations by subverting their goals and objectives to serve the insurgent cause. The insurgents strive to create situations that cause HN police and military forces to overreact in a manner that adversely affects the populace. SF units, with assistance from assigned and attached military police (MP) and CI personnel, advise and assist HN forces engaged in urban area operations. The purpose of these operations is to eliminate the centralized direction and control of the insurgent organization, create insurgent disunity, and destroy the insurgent infrastructure that threatens the HN government. When military forces reinforce police in an urban area, they must closely control and coordinate their operations. By doing so, they minimize collateral damage and prevent hostile propaganda victories that occur when U.S. or HN military forces overreact to insurgent actions. Therefore, the need for PSYOP and CA support greatly increases in urban areas.

POPULACE AND RESOURCES CONTROL

A-47. SF personnel provide advisory assistance in the PRC area as determined by the local situation. Among the considerations are attitudes of the populace; concept, techniques, and control measures of the program; DC operations; and forgiveness and rehabilitation.

ATTITUDES

A-48. Most of the population of any given target area is initially unresponsive to the efforts of the incumbent government or the insurgents. In some societies, there may be a traditional distrust of the government and dissatisfaction with social and economic conditions. However, the population may not have any inclination to revolt. In other societies, a distrust of any influence from "outside" sources may exist. In most instances, the general desire of most of the public is to be left alone to earn a livelihood and to conduct its normal affairs. An effective PSYOP program can exploit this desire for normalcy and direct popular feeling against the insurgents.

A-49. The advocates of revolutionary warfare may be a very small but capable and active segment of the population. Only a small minority of the population may have actively participated in or supported the initial efforts of the insurgents. The forces of the government and its adherents usually represent a countering minority. It includes government officials, civil servants, professional military and police units, leading politicians, the wealthy, and managers of industry, commerce, and banking firms.

A-50. Most of the population remains uncommitted. The insurgents have to persuade or force the population into active or passive support of their goals. The struggle is, therefore, not over terrain. It is a struggle for the support of the populace. If the insurgents win popular support among the majority of the populace, the military successes of the HN government are irrelevant.

CONCEPT

A-51. The design of the PRC program complements and supports the other IDAD programs by providing a secure environment in which to administer these programs. The PRC goals are to—
- Sever the supporting relationship between the population and the insurgents.
- Detect and neutralize the insurgent apparatus and activities in the community.
- Provide a secure physical and psychological environment for the population.

A-52. The HN security forces have primary responsibility for PRC operations. Since civilian communities usually have some system of law and order, a logical approach is to build on the existing law enforcement structure. Some developing countries use paramilitary forces to help civil police in PRC. If a law enforcement system does not exist in the AO, or if the existing structure is corrupt, inept, or compromised, the SF unit may have to help the HN organize, train, and develop a capable police force. When insurgent activities exceed the capabilities of the police and their supporting paramilitary forces, HN regular military forces may have to augment the police. Since the population is more likely to accept control measures enforced by HN personnel than by forces of an outside nation, U.S. forces will normally participate in PRC operations only when the situation is clearly beyond the capabilities of the HN security forces and only when U.S. assistance is requested.

TECHNIQUES

A-53. Intelligence procedures and PSYOP apply to SF in PRC operations. The following paragraphs discuss these procedures and PSYOP.

Intelligence Procedures

A-54. Intelligence must be coordinated at all levels. Intelligence procedures must provide a high degree of penetration of the target, constant pressure, collection of information, and rapid dissemination of intelligence. These procedures allow a quick response by PRC forces. PRC intelligence requirements form a significant part of the overall intelligence effort.

Psychological Operations

A-55. PSYOP are essential to the success of PRC. For maximum effectiveness, SF Soldiers direct a strong PSYOP effort toward the families of the insurgents and their popular-support base. The PSYOP aspect of the PRC program tries to make the imposition of control more palatable to the people by relating the necessity of controls to their safety and well-being. PSYOP efforts also try to create a favorable national or local government image and counter the effects of the insurgent propaganda effort.

CONTROL MEASURES

A-56. SF can advise and assist HN forces in developing and implementing various control measures. The following paragraphs discuss PRC measures.

Security Forces

A-57. Police and other security forces use PRC measures to deprive the insurgent of support and to identify and locate members of his infrastructure. Appropriate PSYOP help make these measures more acceptable to the population by explaining their need. The government informs the population that the PRC measures may cause an inconvenience but are necessary because of the actions of the insurgents.

Restrictions

A-58. Rights on the legality of detention or imprisonment of personnel (for example, habeas corpus) may be temporarily suspended. This measure must be taken as a last resort since it may provide the insurgents with an effective propaganda theme. PRC measures can also include the following:

- Curfews or blackouts.
- Travel restrictions.
- Restricted residential areas, such as protected villages or resettlement areas.
- Registration and pass systems.
- Control of sensitive items (resources control) of critical supplies, such as weapons, food, and fuel.
- Checkpoints, searches, and roadblocks.
- Surveillance, censorship, and press control.
- Restriction of activity that applies to selected groups (labor unions, political groups, and so on).

Deterrents

A-59. Many law enforcement systems have Department of the Army procedures in PRC. They include roadblocks and checkpoints; raids, searches, and screening operations; and mob and riot control. An established reaction force (police or paramilitary personnel) executes these actions, as necessary, and exploits insurgent contacts.

Legal Considerations

A-60. The legality of these methods and their impact on the populace govern all restrictions, controls, and Department of the Army measures. In countries where government authorities do not have wide latitude in controlling the population, special or emergency legislation must be enacted. This emergency legislation may include a form of martial law permitting government forces to search without warrant, to detain without bringing formal charges, and to execute other similar actions.

DISLOCATED CIVILIAN OPERATIONS

A-61. DC operations are a special category of PRC. The goal of this combat support task is to minimize civilian interference with military operations and to protect civilians from military operations. FM 3-05.40 covers DC operations in depth. The SF unit may advise and assist HN forces supporting DC operations by—

- Estimating the number of DCs, their points of origin, and their anticipated direction of movement.
- Planning movement control measures, emergency care, and evacuation of DCs.
- Coordinating with military forces for transportation, MP support, MI, screening, interrogation, and medical activities, as needed.
- Helping them to establish, supervise, and operate DC camps.
- Helping resettle or return DCs to their homes IAW U.S. and HN policy and goals.

FORGIVENESS AND REHABILITATION

A-62. Amnesty, pardon, rehabilitation, and reeducation actions form a distinct and important part of the PRC program. The major aim of this program is to secure the support of the people. To get this support, disaffected members of the population must be able to revert to supporting the government without undue fear of punishment for previous antigovernment acts. Rehabilitation of former insurgent supporters can be through a progressive rehabilitation program. PSYOP forces can actively exploit such programs and greatly increase their effectiveness.

Mission Handoff Procedures

This appendix provides a timeline, or checklist, for a mission handoff. Figure B-1, pages B-3 and B-4, lists in chronological order the tasks the Special Forces operational detachment (SFOD) performs for a mission handoff.

During long-term FID operations, the SF commander may elect to replace an SFOD for various reasons. Mission handoff is the process of passing an ongoing mission from one unit to another with no discernible loss of continuity. It is based on a 179-day requirement and involves two SFODs.

PROCEDURES

B-1. The overall authority for the handoff and assumption of command lies with the commander ordering the change. The authority for determining the handoff process lies with the incoming commander since he will assume responsibility for the mission. This changeover process may affect the conditions under which the mission will continue.

B-2. The outgoing commander advises the incoming commander on the tentative handoff process and the assumption of the mission directly or through a liaison. If this advice conflicts with the mission statement or the incoming commander's desires and the conflict cannot be resolved with the authority established for the incoming commander, the commander ordering the relief resolves the issue.

B-3. As a rule, the commander ordering the change does not automatically place the outgoing SFOD under the incoming SFOD OPCON during the changeover process. Although this procedure would present a clear and easily defined solution to establishing the incoming commander's authority, it is not the most effective control for U.S. forces should hostile contact occur during the process.

B-4. If the incoming SFOD or the HN unit it advises is in direct fire contact with the insurgents during the handoff, the SFOD immediately notifies the higher HQ ordering the exchange. If the incoming SFOD commander has not assumed responsibility, his SFOD immediately comes under OPCON of the outgoing SFOD and is absorbed into that SFOD position. The outgoing SFOD commander and his HN counterpart will control the battle. If the outgoing SFOD commander has passed responsibility to the incoming SFOD commander, the outgoing SFOD comes under the OPCON of the incoming SFOD, and the HN unit coordinates its movements with the new SFOD.

CONSIDERATIONS

B-5. The incoming and outgoing SFOD commanders must consider eight factors:

- *Mission.* The incoming SFOD commander must make a detailed study of the SFOD mission statement and understand the present mission tasks and the implied mission tasks. The mission may also require a unit with additional skill sets such as advanced special operations, direct action, or water operations. Knowing the mission, commander's concept of the mission, commander's critical information requirements (CCIR), priority intelligence requirements (PIRs), and information requirements (IRs) will help him understand the mission. After a complete in-depth study of the operational area, the incoming SFOD commander should complete the handoff in a manner that allows for continued, uninterrupted mission accomplishment. The changeover must not allow the enemy to gain any operational advantages.

- *Operational area.* The in-country SFOD provides continuous intelligence updates to the SF commander. Original PIRs and IRs were established for the original mission along with operational, strategic, and tactical information. The incoming SFOD must become totally familiar with the ongoing PIRs and IRs, and the upcoming mission PIRs and IRs.

- *Enemy forces.* The incoming SFOD commander must have the latest available intelligence on all enemy forces that affect the mission. This intelligence includes data on terrorists and terrorist-related incidents over the past several months. In addition to the normal intelligence provided to the incoming SFOD commander on a regular basis, the situation calls for a liaison from the outgoing SFOD. OPSEC is critical to prevent the enemy from discovering the impending relief and then exploiting the fluidity of the change and the concentration of U.S. forces.

- *Friendly forces.* To the incoming SFOD, learning about the friendly forces is as important as knowing the enemy situation. The SFOD must be familiar with the C2 structure it will deal with on a daily basis. The SFOD must know all friendly units in adjacent AOs and be aware of the conventional forces units and the capabilities of their mission support base. The SFOD must also be aware of other operations, units, and their capabilities. If possible, the incoming SFOD members should receive biographical data on their counterparts, to include photographs. These data allow SFOD members to familiarize themselves with their counterparts before deployment.

- *HN forces.* The incoming SFOD plans and prepares for a quick and frictionless transition in counterpart relations. However, potential or anticipated friction between the HN unit and the incoming SFOD may cause the relief to take place more slowly than desired. Therefore, the incoming and outgoing SFODs need a period of overlap to allow for in-country, face-to-face contact with their counterparts before the mission handoff. Continued execution of the mission must be achieved within the capabilities of the SFODs, the HN unit, and the available supporting assets. If U.S. combat support units are to be relieved, the relief should occur after the relief of the SFODs they support.

- *Civilian populace.* The incoming SFOD must do an in-depth area study, giving close attention to local problems. Popular support for U.S. activities taking place within the AO may directly influence changes in the mission statement. The outgoing SFOD must provide this critical information and describe in detail all completed civic action projects and those that are underway. The incoming SFOD must understand the functioning of the HN government and the status of any international civilian or government agencies involved in, or influencing, the situation in its AO.

- *Terrain and weather.* Some handoff operations may require the SFODs to move by foot into and out of the AO. The outgoing SFOD plans and reconnoiters the routes used for infiltrating the incoming SFOD and those used for its exfiltration. These routes must provide the best possible cover and concealment. If possible, the SFODs make this exchange during darkness or inclement weather.

- *Time.* The depth and dispersion of units and the number of operations conducted will determine the time required to exchange SFODs. There must be an overlap period to allow the incoming SFOD to become familiar with the AO and to establish rapport between the SFOD personnel and their HN counterparts. However, the handoff operation must take place as quickly as possible. The longer the operation takes, the more the SF personnel in the AO become a vulnerable and lucrative target for the insurgents. A quickly executed relief will reduce the time available to the enemy to strike before the incoming SFOD has time to consolidate its position. The SFOD should not sacrifice continued and uninterrupted execution of ongoing operations for speed. The incoming SFOD needs to have enough time to observe training techniques and procedures and to conduct debriefing on lessons learned.

SFOD 945		SFOD 932	
Day	Requirements	Day	Requirements
1	Following planning and preparation, the SFOD (less advance party) **deploys to the AO**. The SFOD deploys with all personnel and equipment required to perform the assigned mission. If this is a first deployment to the AO, the SFOD deploys a site survey team (SFOD members) to coordinate all training with the HN unit and the U.S. Embassy. If there is an SFOD in-country, the incoming replacement SFOD deploys an advance party to coordinate with the deployed SFOD and counterparts. **Area assessment begins** the minute the SFOD members arrive in-country. The SFOD sends the information it gathers from the area assessment to the incoming SFOD through the monthly intelligence summaries (INTSUMs). The SFOD sends timely information at any time, not just through a scheduled monthly INTSUM. As soon as the SFOD arrives, it **establishes communications links** with the higher in-country C2 element. The SFOD also establishes a communications link with the SF commander who has overall authority to order the handoff. This link becomes the information and intelligence link between the in-country SFOD and the incoming SFOD. This link must be maintained and monitored IAW prescribed communications schedules. **Training of the HN begins** as soon as the SFOD is settled.	179	The incoming SFOD **receives notification that it will replace SFOD 945 in 179 days**. The SFOD starts its premission planning. The SFOD **coordinates for its deployment** into the AO. **Leaves, common tasks training, and range qualifications take place** immediately after the mission notification. The SFOD starts training for the mission assigned. The SFOD must use this time wisely. Support for the upcoming mission must come from all levels. The SFOD must **complete certification and validation** before deployment. The SFOD members perform this training as soon as possible to give themselves ample time to heal any sustained injuries. A **review of personnel files** must take place. Any SFOD members considered for career progression schooling must be taken into account. Every effort is made to send them to these schools as soon as possible. If an SFOD member is scheduled for one of these schools during the MTT, a replacement is nominated.
30	The SFOD sends training reports at least every 30 days. These reports indicate how the HN unit is responding to the in-country SFOD training program. The incoming SFOD uses these reports to modify its training programs and schedules. The SFOD sends an INTSUM at least every 30 days or as the military or political situations change. INTSUMs are not restricted to monthly transmissions only.	90	The SFOD receives formal mission notification. All travel arrangements are finalized; the in-country SFOD is notified.
45	Command inspections performed by the SAO are a vital part of an MTT. They give the commander a chance to see firsthand what an SFOD has accomplished to date. The inspection will ensure that the HN is looking after the welfare of the SFOD, and any problems with the HN are corrected immediately. If possible, one member from the incoming SFOD accompanies the commander.	85	The formal SFOD train-up program begins. All other support requirements must stop; the SFOD mission must take priority.
60	Training reports, same as day 30. INTSUM, same as day 30.	80	The SFOD senior medical NCO begins screening the members' records. He ensures all personnel have physicals and their shots are up-to-date. He obtains and cross-matches their blood types within the team. The SFOD prepares initial shortage lists and sends them to the battalion S-4. If the SFOD requires special items, it requests them as soon as possible. An initial preparation of replacements for overseas movement (POR) is scheduled.
89	The detailed intelligence report contains more than the monthly INTSUM. This report becomes a major part of the incoming SFOD mission planning process. The incoming SFOD receives this report the same day it receives its mission notification and starts its detailed planning.		

Figure B-1. SFOD 945 hands off to SFOD 932

SFOD 945		SFOD 932	
Day	Requirements	Day	Requirements
90	Training reports, same as day 30. INTSUM, same as day 30. Mid-tour leaves are programmed into the training schedule. These leaves must be staggered so that they do not interfere with training. In-country leaves should be considered.	70	Final leaves are granted to SFOD members. They ensure their personal affairs are in order.
100	Command inspection.	60 to 50	A team from the incoming SFOD may conduct an in-country coordination. The SFOD commander ensures only his personnel are on this team.
110	The SFOD finalizes its travel arrangements. The initial arrangements were made the day the SFOD deployed.	40	The SFOD prepares its final shortage lists and sends them to the battalion S-4. The SFOD picks up specialized items from the S-4.
119	The incoming SFOD has had its formal mission notification and has started its mission preparation.	30	The SFOD medics complete a final medical screening. Members receive their shots at this time. All injuries sustained during the certification and validation should be healed. Personnel who require more time to recover may be replaced. The SFOD-appointed Class A agent draws the advance. Ordering officers are appointed and receive their briefing from finance.
120	Training reports, same as day 30. INTSUM, same as day 30.		
135	Command inspection.		
150	Training reports, same as day 30. INTSUM, same as day 30.		
165	This period is the most critical phase of the mission. The SFOD members who were in the advance party depart. The incoming SFOD advance party quickly meshes with the remaining in-country SFOD members and their HN counterparts. They establish rapport and begin the next phase of training. The in-country SFOD must have completed all training by this time.	25	All SFOD personnel draw their advances. This action allows the correction of any problems before deployment.
179	The in-country SFOD (minus its already departed members) departs. Army regulation requires temporary duty (TDY) personnel leave not later than (NLT) this date.	20	The SFOD members' dependents (families) receive a briefing. Every effort is made to answer all questions that would not create a security risk for the deploying SFOD. Dependents are provided with a POC in the event of problems.
		18	The SFOD palletizes all its equipment and personal gear. The SFODB team must ensure the SFOD has total cooperation from the SFODC S-4.
		16	The SFOD establishes a communications link.
		14 to 2	The advance party members deploy to the AO. They meet with the in-country SFOD members, establish rapport with their HN counterparts, and conduct all necessary coordination.
		1	The incoming SFOD (less its advance party) deploys to AO. The mission handoff is completed.

Figure B-1. SFOD 945 hands off to SFOD 932 (continued)

Postmission Debriefing Procedures

The SFOD commander conducts a debriefing that provides an overview of the mission, military geography, political parties, military forces, insurgents, security forces, underground, targets, health and sanitation practices of the populace, and evasion and recovery (E&R). Figure C-1, pages C-2 through C-6, depicts a guide for conducting a debriefing.

POSTMISSION DEBRIEFING

C-1.   Redeployment is not the end of the mission. Upon arrival at the redeployment location, the SFOD undergoes an extensive debriefing. The battalion S-2 officer organizes and conducts the debriefing, subject to unit SOP.

C-2.   The S-2 coordinates with higher-level intelligence organizations to take part in the SFOD debriefing, particularly if other organizations tasked the SFOD to obtain information. All deployed personnel, to include attachments, must be available for the debriefing.

DOCUMENTATION

C-3.   After the debriefing, the SF team leader, with the assistance of other members of the team and attachments, prepares two documents. The unit historian prepares a third document.

C-4.   The first is an after action review (AAR). The AAR states the who, what, when, where, and how of the operation. It is a permanent record of the major activities of the team from isolation to debriefing. As such, it is an extremely important template on which past missions may be compared and future missions planned. The AAR is normally submitted through command channels to the group commander NLT 48 hours after an SF team has been debriefed. The intelligence and operations officers at each echelon keep copies of SF team AARs. If applicable, the unit historian also reviews the AAR and prepares a draft report for entry into the unit history.

C-5.   Shortly after completion of the AAR, or simultaneously with its submission, the team leader submits a report of lessons learned. This report is the team leader's reflection on his most recent operation and his recommendation for the future. One method is to organize the lessons according to the six warfighting functions: movement and maneuver, intelligence, fire support, sustainment, C2, and protection. It addresses what worked and what did not work on the operation, why it did or did not work, and what changes or substitutions are needed for existing TTP in the unit.

C-6.   The unit historian reviews the report of lessons learned and then completes the unit history for the operation, subject to the commander's approval. The historian issues an official historical report of the operation in classified and unclassified versions, as appropriate, within 90 days after the completion of the operation.

MISSION
• Brief statement of mission by SFOD commander.

EXECUTION
• Brief statement of the concept of operation developed before the deployment.
• Statement of method of operation accomplished during the operation, to include deployment, routes, activity in HN areas, and redeployment.
• Uniforms and equipment used.
• Weapons, demolitions, and ammunition used and results.
• Communications equipment used and results.
• Casualties (friendly and/or enemy) sustained and disposition of bodies of those killed in action (KIA).
• Friendly contacts established, to include descriptions, locations, circumstances, and results.

MILITARY GEOGRAPHY
• Geographic name, Universal Transverse Mercator or geographic coordinates, and locations.
• Boundaries (north, south, east, and west).
• Distance and direction to nearest major cultural feature.
• Terrain.
▪ What type of terrain is dominant in this area?
▪ What natural and cultivated vegetation is present in the area?
▪ What is the density and disposition of natural vegetation?
▪ What is the approximate degree of slope?
• What natural obstacles to movement did you observe and what are their locations?
• What natural or man-made drainage features are in the area?
▪ Direction of flow.
▪ Speed.
▪ Depth.
▪ Type of bed.
• What is the physical layout of rural and urban settlements?
• What is the layout of various houses within the area?
• How would you describe any potential landing zones (LZs) or drop zones (DZs)?
• How would you describe any beach landing sites, if applicable?
• How would you describe any areas suitable for cache sites and what are their locations?
• People.
▪ What major ethnic groups or tribes populate each area?
▪ What was (or is) their attitude toward other ethnic groups or tribes in the area?
▪ What is the principal religion of the area and how is it practiced?
o Influence on people.
o Religious holidays.
o Taboos.
o Conflicts in religions.
▪ How would you describe the average citizen of the area (height, weight, hair color, characteristics)?
▪ How do the people of this area dress compared with other areas?
▪ What type clothing, footwear, ornaments, and jewelry do they wear?
▪ What symbolism is attached to certain items of jewelry and/or ornaments?
▪ What are the local traditions, customs, and practices?
o Between males and females?
o Between young and old?
o Toward marriage, birth, and death?
o Between the populace and local officials?
▪ What is the ordinary diet of the people?
▪ What was the attitude of the populace toward you and the HN forces with you?
▪ What was the general feeling and attitude of the populace and the HN troops toward the government and leaders, government policies, and general conditions within the country?
▪ What was the general feeling of the populace toward the United States, its policies, and involvement with other nations?

Figure C-1. Postmission debriefing guide

MILITARY GEOGRAPHY (continued)
• How did the populace cooperate with USSF?
• What is the approximate wage and economic status of the average citizen?
• What formal and informal educational practices did you observe?
• What is the state of health and well-being of the people in this area?
• Did the populace in this area speak the national language differently from others in the country? If so, how?
• What percentage of the populace and the indigenous forces speak English or other foreign languages?
• Were you approached or questioned by some member of the populace about the USSF or your mission? If so, describe in detail. Give names, if possible.

POLITICAL PARTIES (Major and Minor Parties)
• Leaders.
• Policies.
• Influence on government.
• Influence on the people.
• Foreign influence.
▪ Ethnic and/or ideological.
▪ Regional.
▪ International.
▪ Stability, strength, and weaknesses.

MILITARY
• Friendly forces.
• Disposition.
• Composition, identification, and strength.
• Organization, armament, and equipment.
• Degree of training, morale, and combat effectiveness.
• Mission.
• Leadership and capabilities of officers and NCOs compared with those of the United States.
• Logistics.
• Maintenance problems with weapons and equipment.
• Methods of resupply and their effectiveness.
• Psychological strengths and weaknesses.
• Relationship between HN military forces, the populace, and other forces (paramilitary, police, and CSDF).
• Influence on local populace.
• Recommendation for these forces (military and/or paramilitary) for UW contact.

INSURGENT FORCES
• Disposition.
• Composition, identification, and strength.
• Organization, armament, and equipment.
• Degree of training, morale, and combat effectiveness.
• Mission.
• Leadership capabilities.
• Logistics.
• Maintenance problems with weapons and equipment.
• Method of resupply and its effectiveness.
• Psychological strengths and weaknesses.
• Relationship between insurgent forces, your SFOD, and the populace.
• Influence on local populace.

POLICE AND SECURITY FORCES (Friendly and Enemy)
• Disposition, strengths, and location.
• Organization, armament, and equipment.
• Logistics.
• Motivation, reliability, and degree of training.

Figure C-1. Postmission debriefing guide (continued)

POLICE AND SECURITY FORCES (Friendly and Enemy) (continued)
• Psychological strengths and weaknesses.
• Relationship with the government and local populace.
AUXILIARY AND UNDERGROUND
• Disposition, strength, and degree of organization.
• Morale and general effectiveness.
• Motivation and reliability.
• Support.
▪ Logistics.
▪ Intelligence.
TARGETS **Describe the area:**
• Rail system.
▪ General route.
▪ Importance to the local and general area.
▪ Bridges, tunnels, curves, and steep grades.
▪ Bypass possibilities.
▪ Key junctions, switching points, and power sources.
▪ Location of maintenance crews who keep the system operational during periods of large-scale interdiction.
▪ Security.
• Telecommunications system.
▪ Location and description of routes, lines, and cables.
▪ Location of power sources.
▪ Location and capacity of switchboards.
▪ Critical points.
▪ Importance to the local general area.
▪ Capabilities of maintenance crews to keep the system operating at a minimum.
▪ Security.
• Petroleum, oils, and lubricants (POL) storage and processing facilities.
▪ Location.
▪ Capacity of storage facilities.
▪ Equipment used for the production of POL.
▪ Power source.
▪ Types and quantities of POL manufactured.
▪ Methods of transportation and distribution.
o Rail.
o Truck.
o Ship.
o Air.
▪ Pipeline routes and pumping station capacities.
▪ Security.
• Electrical power system.
▪ Location and description of power stations.
▪ Principal power lines and transformers.
▪ Location of maintenance crews, facilities, and reaction time.
▪ Critical points.
▪ Capacity (kilowatts).
▪ Principal users.
▪ Security.
• Military installations and depots.
▪ Size.
▪ Activity.
▪ Location.
▪ Units.

Figure C-1. Postmission debriefing guide (continued)

TARGETS (continued)
Describe the area:

- Equipment.
- Reaction time.
- Security.
- Highway and road system.
  - Name and number.
  - Type of surface, width, and condition.
  - Location of bridges, tunnels, curves, and steep grades.
  - Bypass possibilities.
  - Traffic density.
  - Location of maintenance crews, facilities, and reaction time.
  - Security.
- Inland waterways and canals.
  - Name and number.
  - Width, depth, and type of bed.
  - Direction and speed of flow.
  - Location of dams and locks, their power source, and other traffic obstructions.
  - Location and descriptions of administrative, control, maintenance crew, facilities, and reaction crew.
  - Location and description of navigational aids.
- Natural and synthetic gas system.
  - Location and capacity of wells and pipelines.
  - Storage facilities and capacity.
  - Critical points.
  - Maintenance crews, facilities, and reaction time.
  - Principal users.
  - Security.
- Industrial facilities.
  - Capabilities of plants to convert their facilities in wartime to the production of essential military materials.
  - Type of facilities.
  - Power sources.
  - Locations.
  - Sources of raw materials.
  - Number of employees.
  - Disposition of products.
  - General working conditions.
  - Critical points.
  - Security.

HEALTH AND SANITATION

- To what degree does hunting and fishing contribute to the local diet?
- What cash crops are raised in the area?
- What domestic and wild animals are present?
- What animal diseases are present?
- What is the availability and quality of water in populated and unpopulated areas?
- What systems are used for sewage disposal?
- What sanitation practices did you observe in the populated and unpopulated areas?
- What are the most common human illnesses and how are they controlled?

EVASION AND RECOVERY

- From which element of the populace is assistance most likely?
- Would you recommend any safe houses or areas for E&R purposes?
- What type shelters were used?
- Were fires small and smokeless?
- Were shelters adequate?
- Was food properly prepared?

Figure C-1. Postmission debriefing guide (continued)

EVASION AND RECOVERY (continued)
• Were camp sites well chosen?
• Were camp sites and trails sterilized after movement to a new one?
• What edible wild plants are found in the area?
MISCELLANEOUS
• Weather.
▪ Wind speed and direction.
▪ Temperature.
▪ Effect on personnel and equipment.
• Problems encountered.

Figure C-1. Postmission debriefing guide (continued)

Appendix D

Site Survey Procedures

The site survey checklist is a tool used by the site survey team to help them answer questions identified by the SF unit during their preparation for deployment. The checklist shown in Figure D-1, pages D-4 and D-5, is a guide and not meant to be all-inclusive. The checklist can be modified as needed. The SF unit can modify it to aid the site survey team in acquiring needed information for planning before their deployment.

SITE SURVEY TEAM MISSION

D-1. The mission of the site survey team is to report accurately to its parent unit the existing HN mission, enemy, terrain and weather, troops and support available, time available, civil considerations (METT-TC) conditions. It also establishes in-country C2 systems and logistics relationships for the follow-on unit mission execution and coordinates the in-country reception of the main body.

SITE SURVEY TEAM PROCEDURES

D-2. Before departure, the site survey team conducts predeployment activities to include—

- Obtaining, through the battalion S-2 and S-3, the required travel documents (visas and passports) and a copy of the country clearance message sent by the U.S. Embassy, if required.
- Ensuring all site survey team members' medical and immunization records are current.
- Conducting predeployment finance operations.
- Receiving the Security Assistance Training Management Organization (SATMO) briefing (if applicable).
- Obtaining designated fund cites.
- Confirming, with the U.S. HN team, that all agencies concerned with the site survey have been briefed on the team's itinerary and are available for coordination.
- Receiving and updating the threat briefing and reviewing the ROE and status-of-forces agreement (SOFA) (if any).
- Conducting a mission analysis and briefback IAW unit SOP. The team tailors its mission analysis and briefback to the site survey mission.

D-3. Upon arrival in the HN, the team processes through customs, notifies the SAO of its arrival and status, and requests an updated threat briefing. The survey team must be ready to brief the mission and program of instruction (POI) to the SAO for approval and/or modification.

D-4. The survey team commander and S-3 establish the command relationship with the next-higher in-country U.S. commander if he is not in the team's normal chain of command. The team commander also briefs the next-higher in-country U.S. commander on the planned execution of the survey and the required preparations for the main body.

D-5. The survey team commander also obtains any additional guidance from the higher in-country U.S. commander for the follow-on forces' mission execution. As a minimum, this guidance includes confirmation of the ROE, E&R support, and the limitations on relationships with HN counterparts. The survey team commander discusses the following areas with the SAO:

- Training objectives.
- Terms of reference.
- Political situation.

- Social customs.
- Guidelines for official and personal associations with foreign personnel.
- Currency control.
- Procedures for obtaining intelligence support from the next higher in-country U.S. commander.
- Administrative support.
- Legal status in relation to the foreign country (SOFA).
- Procedures for obtaining logistics from the next-higher in-country U.S. commander.

D-6. The team commander confirms or establishes communications and reporting procedures between the next-higher in-country U.S. commander, the survey team, and the follow-on SF units still in mission preparation. The team commander must also identify the availability of communications equipment needed to support the mission.

D-7. The team commander confirms or establishes procedures for obtaining logistics from the next-higher in-country U.S. commander. He identifies a POC at the Country Team crisis management element or at the emergency operations center (EOC) of the U.S. military staff. The POC then informs the SF unit of necessary actions during increased threat or emergencies that require evacuation of U.S. personnel from the HN. The team commander establishes the procedures to obtain intelligence support from the higher in-country commander or other U.S. agencies.

D-8. The survey team establishes direct working relationships with its next-higher in-country or out-of-country support element. The survey team—

- Identifies the supporting element location.
- Contacts the supporting element to determine the limitations of the available support and the expected reaction time between the initiation of the support request and its fulfillment.
- Requests support for the in-country reception of the main body IAW the requirements in the survey team OPORD.
- Confirms or establishes communications procedures among the supporting element, the survey team, and the follow-on SF unit still in mission preparation. It identifies, as a minimum, alternate and emergency communications procedures for C2, all available logistics, and medical evacuation (MEDEVAC).
- Reports the established communications support-requesting procedures for the follow-on SF unit.

D-9. The survey team establishes procedures to promote interagency cooperation and synchronize operations. The team—

- Identifies the location of the concerned HN or U.S. agency.
- Contacts the concerned agency to establish initial coordination.
- Exchanges information and intelligence.
- Confirms or establishes communications procedures.
- Confirms or establishes other coordination protocols, as necessary.
- Reports the newly established or changed procedures for inclusion into the follow-on SF plans for mission execution.

D-10. The survey team commander and/or specified subordinates establish direct working relationships and rapport with the HN unit commander. The survey team—

- Briefs the HN commander on the SF unit survey mission and the restrictions and limitations imposed on the SF unit by the higher U.S. commander.
- Assures the HN commander that his assistance is needed to develop the tentative objectives for training and/or advisory assistance.
- Deduces or solicits the HN commander's actual estimate of his unit capabilities and perceived training and/or advisory assistance and material requirements. They discuss training plans, current training status and/or needs, units available for training, and training facilities.

D-11. The survey team obtains the HN commander's approval of the plan. The team also requests linkup with the counterpart under the mutual supervision of the HN commander and the survey team commander.

D-12. The team does not make any promises (or statements that could be construed as promises) to the HN commander about commitments to provide assistance or fulfill material requirements. In particular, the survey team does not—

- Make any comment to host government on possible availability of USG resources in any form.
- Provide any kind of independent assessment or confirmation of the external threat, as perceived by the HN.
- Provide advice on tactics, doctrine, basing, combat planning, or operations.

D-13. The survey team analyzes the HN unit status to determine HN requirements for training and/or advisory assistance. The team—

- Collects enough information to confirm the validity of current intelligence and selects tentative training and/or advisory assistance COAs.
- Prepares written estimates for training and/or advisory assistance COAs that are prioritized in order of desirability.
- Determines the unit location and its effects on the populace.
- Collects and analyzes all information affecting force protection.
- Determines the HN unit's existing logistics and maintenance support shortfalls and capabilities.
- Determines the compatibility of recommended equipment with that in the HN inventory.

D-14. The survey team helps the HN unit prepare facilities (training, security, and administrative) for the execution of the mission. The survey team inspects the HN facilities the SF unit members and their counterparts will use during the mission. At this time, it identifies any deficiencies that will prevent the execution of the tentatively selected training and/or advisory assistance COAs. After the inspection, the survey team commander recommends to the HN commander the most desirable COAs to correct any deficiencies found.

D-15. The survey team commander recommends to the HN commander the most desirable COAs, emphasizing how to achieve the desired training and/or advisory assistance objectives. The survey team commander—

- Ensures the HN commander understands the desired COAs are still tentative (contingent on the U.S. commander's decision).
- Ensures the higher in-country U.S. commander is informed of significant findings in the survey of the HN unit.
- Selects the COAs to be recommended to the follow-on SF units, after obtaining input from the HN commander.

D-16. The survey team ensures its security at all times, according to the latest threat assessment. The team—

- Fortifies its positions (quarters, communications, medical, and command) within the available means, keeping in mind the requirement to maintain low visibility.
- Establishes and maintains an internal alert plan.
- Organizes and maintains an internal guard system with at least one member who is awake and knows the location of all team members. The guard is ready to react to emergencies by following the alert plan and starting defensive actions according to established ROE and E&R procedures.
- Maintains communications with all team members outside the immediate area occupied by the team's main body.

D-17. Before departing from the HN, the survey team again visits all concerned U.S. and HN staff agencies to clarify any unresolved problem areas.

Security Assistance Organization
S-2
(1) Intelligence briefing. (2) Threat briefing. (3) Maps and photos of the area. (4) Weather forecast data. (5) Restricted and off-limits areas. (6) Local populace (attitudes, customs, and dangers).
S-3

(1) Initial coordination.
  (a) Tentative training plans.
  (b) Aviation support tentatively available (hours and type of aircraft).
  (c) HN plans (tentative).
  (d) Problem areas.
  (e) Evasion plan of action (EPA)-related directives, guidance, plans, or orders.
(2) POC, phone number list, communications requirements, and systems used.

S-4
(1) Transportation requirements. (2) Special equipment requirements. (3) Other support requirements. (4) Construction equipment and supply requirements.
Host Unit
Commander
(1) Training plan. (2) Current training status. (3) Units available for training. (4) C2. (5) Additional training desires. (6) Unit policies.
S-2
(1) Local civilians. (2) Security policies and problems. (3) Populace control requirements (identification [ID] cards/passes).
S-3

(1) Training plan.
(2) Support available.
  (a) Ammunition.
  (b) Weapons.
  (c) Vehicles.
  (d) Aircraft/air items.
  (e) Facilities:
    • Training areas.
    • Classrooms.
    • Ranges.

Figure D-1. Suggested site survey checklist

S-3 (continued)
• Training aids.
• Special equipment.
(3) Unit equipment.
(4) LZs and DZs in the area.
(5) Maps.
(6) Rations for field training.
(7) Daily training schedules and status reports.
(8) POC for training problems.
(9) Holidays and unit requirements that may interfere with training.
(10) Medical and dental support.
(11) Communications capabilities.
(12) HN activities.
S-4
(1) Detachment facilities.
(a) Barracks.
(b) Drinking water.
(c) Messing facilities.
(d) Secure storage areas.
(e) Electrical power supply.
(2) Fuel supply.
(3) Rations.
(4) Transportation.
(5) Lumber and materials for training aids.
(6) Special equipment.
(7) Ammunition.
(8) Availability of construction equipment/tools and supplies.

Figure D-1. Suggested site survey checklist (continued)

This page intentionally left blank.

Legal Considerations

FID operations must be conducted IAW international law and U.S. domestic law. U.S. SA and arms transfers programs are subject to specific congressional authorization, appropriation, and oversight. Commanders and other FID planners must consult with their legal advisors to ensure they conduct operations IAW ever-changing U.S. legislation and policy.

In general, legal considerations on the international level center on the question of describing the conflict in the HN as international or internal (insurgency). Legal considerations for the United States mainly involve using the proper funds for the type of mission being conducted. Additional country issues and specific U.S. legislation must also be considered.

INTERNATIONAL LEGAL CONSIDERATIONS

E-1. Under international law, armed conflicts fall into two broad areas. These areas are those of an international character and those not of an international character.

INTERNATIONAL CONFLICTS

E-2. A declaration of war and an invasion of one country by the armed forces of another clearly result in international conflict. The definition of an international conflict is broader, however. As a rule, if the combat effects of a conflict go beyond a nation's boundaries and seriously affect other countries, the conflict is international. All the customary laws of war on hostilities between states govern international armed conflicts. The 1949 Geneva Conventions and all other treaties that make up the laws of war also apply. As a practical matter, an important concern of the Soldier fighting in this type of war is his right to prisoner of war (PW) status if captured.

NONINTERNATIONAL CONFLICTS

E-3. Noninternational conflicts are typically called insurgencies. Clandestine forces usually engage in hostilities. Their purpose is not to hold fixed territory or to engage government troops in direct combat but to wage a guerrilla-type war. In this war, they can lose themselves in the civilian populace by posing as noncombatants. Insurgents, therefore, are organized bodies of people who, for public political purposes, are in a state of armed hostility against the established government. An important legal aspect of a noninternational conflict is that captured combatants do not enjoy the rights of PWs. They may be prosecuted as criminals under the laws of the HN. The fact that an insurgent follows the rules of war or is in uniform will not give him PW status under international law. Article 3 of each of the four Geneva Conventions of 1949 provides the primary source of rights and duties of persons involved in noninternational conflicts. Common Article 3 has two parts.

First Part

E-4. The first part provides that persons taking no active part in the hostilities, including members of armed forces who have laid down their arms and those out of combat because of sickness, wounds, detention, or any other cause, shall in all circumstances be treated humanely. Humane treatment specifically excludes—

- Violence to life and person; in particular, murder, mutilation, torture, or any cruel treatment.
- Hostage-taking.

- Outrages upon personal dignity; in particular, humiliating and degrading treatment.
- Passing of sentences and carrying out executions without previous judgment pronounced by a regularly constituted court affording all the judicial guarantees that civilized peoples recognize as vital.

Second Part

E-5. The second part requires collecting and caring for the wounded and sick. Common Article 3 does not grant PW status to insurgents. It does require the government to grant them a fair trial in a regularly constituted court before carrying out the court's sentence after a guilty verdict.

E-6. Common Article 3 incorporates basic human rights. Human rights also include other rights embodied in the phrase "life, liberty, and the pursuit of happiness," such as the right of free speech, freedom of worship, and freedom of the press. U.S. personnel who notice suspected violations of basic human rights must report the facts to their chain of command. Under U.S. law, the President must cut off SA to any country with a documented pattern of human rights abuses.

UNITED STATES LEGAL CONSIDERATIONS

E-7. Funding for FID activities comes from two principal sources: Foreign Assistance Act (FAA) funds appropriated to DOS IAW Section 2151, Title 22, United States Code (22 USC 2151), *Congressional Findings and Declaration of Policy*, and operations and maintenance (O&M) funds appropriated to DOD IAW 10 USC. Congress and the General Accounting Office exercise close oversight to ensure O&M funds are not used for activities that should have been funded through FAA funds.

E-8. Commanders must be able to distinguish FAA-funded activities from DOD-funded activities. Using the wrong funds can violate the Antideficiency Act (Section 1341, Title 31, United States Code [31 USC 1341], *Limitations on Expending and Obligating Amounts*). Antideficiency Act violations are reportable to Congress and carry both civil and criminal penalties.

TITLE 22 PROGRAMS

E-9. Included in Title 22 programs are the FAA and the AECA. The FAA and AECA are discussed in the following paragraphs.

Foreign Assistance Act

E-10. The FAA (22 USC 2151) is the most comprehensive of the statutes dealing with SA. The FAA provides economic, agricultural, medical, disaster relief, and other forms of assistance to developing countries. The FAA also assists foreign countries in fighting internal and external aggression by providing various forms of military assistance upon request (and subject to congressional approval). Despite a large DOD role in providing defense-related articles and services, the DOS controls the FAA. The FAA mandates close coordination and cooperation between DOD and U.S. civilian agencies at all levels of the SA process. Principal programs under the FAA include the following:

- *Foreign Military Financing Program.* This program consolidates three former SA programs: the Foreign Military Sales Financing Program, the Foreign Military Sales Credit Program, and the Military Assistance Program. Although intended as a grant and a loan program, the Foreign Military Financing Program provides the bulk of assistance on a grant basis.
- *International Military Education and Training.* This program authorizes military education and training to military and related civilian personnel of foreign countries, primarily at schools in the United States.
- *Antiterrorism Assistance.* This program provides training to foreign country law enforcement personnel to enhance their ability to deter terrorist activities. Training services furnished under this program cannot take place outside the United States. To the maximum extent possible, U.S. advisory personnel must carry out their duties within the United States.

Arms Export Control Act

E-11. The AECA contains the FMS program. The AECA provides for the transfer of arms and other military equipment, as well as various defense services (such as training) through government-to-government agreements. Under this program, defense articles and services are sold, not given away. The law prohibits personnel providing services under this program from engaging in any duties of a combat nature. This prohibition includes any duties related to training and advising that may engage U.S. personnel in combat activities. Although they may engage any hostile force in self-defense, training teams or personnel should withdraw as soon as possible.

TITLE 10 PROGRAMS

E-12. Included in Title 10 programs are O&M funds and HCA. These programs and their related activities are discussed in the paragraphs below.

Operations and Maintenance Funds

E-13. These funds are appropriated for the support of the U.S. military. DOD has a good deal of discretion in how to spend these general-purpose funds. Under fiscal law principles, DOD cannot spend them for any foreign assistance activity for which Congress has specifically appropriated funds. Some O&M-funded DOD activities are on the periphery of SA programs, and commanders must be alert to the differences.

Coalition Operations

E-14. A mission of DOD is coalition operations—knowing how to fight alongside the armed forces of friendly countries. The U.S. Comptroller General has established the following fiscal law principles on combined training:

- Combined exercises that provide overseas training opportunities for U.S. personnel and support the goals of U.S. coalition operations may use O&M funds despite providing training to HN forces.
- The permissible scope of HN training includes safety, familiarization, and interpretability training.
- Combined exercises assume the involvement of comparably proficient units. O&M funds may not be used to provide the level of training available through SA programs.
- O&M funds are provided for U.S. forces to take advantage of opportunities to train with foreign forces. SA funds are intended for U.S. forces to provide concentrated training for foreign forces.

Special Forces Exception

E-15. The Comptroller General has acknowledged that SF Soldiers have a mission to train foreign forces. SF may train a foreign military force to test their ability to accomplish their mission. The primary goal or benefit must be to test SF training capabilities. Title 10 has been amended expressly to authorize the use of O&M funds to finance SF training with foreign forces (10 USC 2011). This training is permissible as long as it is not comparable to or intended as SA training; that is, the training must be conducted as an SF team and not be long-term.

Humanitarian and Civic Assistance

E-16. HCA projects are among the most effective instruments for dealing with HN conditions conducive to the emergence of insurgencies. Until the fiscal year (FY) 1987 DOD Authorization Act, HCA was not a DOD mission. Instead, HCA was funded as a form of SA undertaken by USAID. DOD authority was limited to HCA provided from DOD assets to USAID on a reimbursable basis or to HCA provided incidental to exercises directed by the JCS. In the Authorization Act, Congress specifically authorized DOD-provided HCA activities. HCA authorities include the following:

- *"de minimis" HCA.* DOD may spend minimal O&M funds for de minimis HCA when unplanned HCA opportunities occur. This term would include a unit doctor's or medic's examination of villagers for a few hours or giving inoculations and issuing some medicines. However, this term would not include the dispatch of a medical team for mass inoculations.

- *Inherent authority.* DOD has an inherent authority to undertake HCA activities that, by chance, create HCA benefits and are carried out to fulfill the training requirements of the unit involved. U.S. medical readiness training is an example.

- *Stevens Amendment.* This amendment authorizes DOD personnel to conduct HCA activities with CJCS and/or combatant-commander-directed OCONUS exercises. The HCA activities can be unrelated to their own training requirements. The amendment was originally a temporary solution that has continued through DOD appropriations.

- *Interagency transactions.* Under the Economy Act, DOD personnel may conduct HCA activities for another federal agency, primarily DOS. Prior arrangements must be made for DOS to reimburse DOD for any costs incurred.

- *Statute.* 10 USC 401 specifically authorizes DOD to provide HCA. HCA is specifically defined as—

  - Medical, dental, and veterinary care provided in rural areas of a country.
  - Construction of rudimentary surface transportation systems.
  - Well drilling and construction of basic sanitation facilities.
  - Rudimentary construction and repair of public facilities.

E-17. The Secretaries of Defense and State must specifically approve in advance HCA rendered pursuant to this authority. Payments are made from O&M funds specifically appropriated for HCA. An important limitation is that HCA may not be provided to any military or paramilitary individual, group, or organization.

LEGAL STATUS

E-18. Usually, anyone present in a foreign nation's territory is subject to its jurisdiction. Jurisdiction is the legal power a sovereign nation has to make and enforce its laws without foreign dictation.

E-19. When a nation's troops enter a friendly foreign country, international law subjects them to the territorial jurisdiction of that nation and any jurisdiction, because of their status, the sending state wishes to exercise. U.S. military forces are always subject to the Uniform Code of Military Justice (UCMJ).

E-20. U.S. policy is to maximize U.S. jurisdiction over the armed forces it may deploy to a foreign nation. The legal status of U.S. forces in a foreign nation is usually defined in one of the following types of international agreements:

- Emergency wartime agreements.
- SAO agreements.
- SOFAs.

E-21. During military emergencies, the United States normally obtained exclusive jurisdiction over its troops in foreign countries. Emergency agreements have normally been short and uncomplicated. The classic examples of these types of agreements are the 1950 Korea, the 1968 Lebanon, and the 1984 Grenada stationing agreements.

E-22. SAO agreements provide a lower level of diplomatic immunity to U.S. troops stationed in countries under these agreements. Each agreement is individually negotiated with the country in question and, therefore, is usually different. Soldiers on TDY in these countries (for example, a FID mission) are usually attached to the SAO and automatically assume the protection accorded those personnel. Agreements of this type normally provide the same diplomatic immunity for anything done in the performance of official duty. Personnel performing a FID mission may come within the scope of the SAO agreement itself or be included by the terms of an SA "contract" entered into between the United States and the HN.

E-23. SOFAs are the most comprehensive type of international agreements. SOFAs are usually used where the United States has stationed many forces for an extended period (Germany and Korea). SOFAs usually provide for a sharing of jurisdiction over U.S. forces with the United States having the primary right to exercise jurisdiction over offenses solely involving—

- U.S. members or property.
- Security of U.S. forces.
- Actions occurring in the performance of official duty.

E-24. U.S. forces performing a FID mission are not automatically immune from HN jurisdiction. Commanders coordinate with their legal advisor to find out the legal status of their personnel and try to obtain any necessary protection if there is no applicable international agreement.

This page intentionally left blank.

Appendix F

Advisor Techniques

The advisor techniques outlined in this appendix apply to the individual advisor and an SF unit in FID operations. In some instances in the past, U.S. advisors were not selected based on language skills or ability to deal effectively with their counterparts. They were selected based on military occupational specialty (MOS) and availability for an overseas hardship tour. The U.S. military services have demonstrated their professional excellence in training foreign personnel and units in technical skills. However, they have not performed well in advising in politico-military matters because of their lack of background, training, and competence in these areas.

APPLICATION

F-1.   Influencing HN military institutions to support a democratic process can only be done with the long-term presence of U.S. military personnel working alongside HN forces. Personnel who arrive for short visits will be treated as visitors and will not penetrate the fabric of the HN culture or its institutions. Although short visits can serve other useful purposes, the long-term presence of U.S. military personnel is required to strengthen HN democratic institutions and convince the HN military institutions to reform. HN officials are not normally confused over moral ground rules; however, because of the dangerous situation confronting the nation, they are convinced they must ignore these ground rules.

F-2.   An advisor must strive to transmit the concept of "democratization" to his counterpart. These concepts are often considered "common sense" or "common decency" and so basic in the United States they are not discussed much in training. The most important mission of an advisor is to enhance the military professionalism of his counterpart. He must influence the HN military and prepare them to deal with the changing environment by emphasizing civilian control over the military and demonstrating the advantages of a democratic system of government.

F-3.   A major cause of an advisor's failure is his inability to maintain a good working relationship with his counterpart. The unsuccessful advisor often fails to understand why his counterparts may not feel the "sense of urgency" that he does. He is unable to realize that his counterpart will remain and continue to fight the enemy long after his tour is over and he returns to the safety and comfort of the United States.

F-4.   The advisor must be aware of the scope and limitations of the principal SA programs authorized by the FAA and AECA. The current ROE for the AO will determine what level of SA personnel may perform any duties of a combatant nature. These include any duties related to training and advising that may engage U.S. personnel in combat activities. The environment plays a big factor in an advisor's role. The following paragraphs address what the advisor needs to know to prepare for his role.

STAGE OF DEVELOPMENT

F-5.   In situations where the HN government may have been in existence only a short time, the administrative machinery may still be developing. The advisor must be aware of such situations and not be overly critical. In an insurgency, the HN government is experiencing major problems. For instance, the money needed for social and economic programs is mostly directed toward security needs. In an ideal situation, the HN government would use this money to cure the society's economic and social ills.

ORGANIZATIONAL MAKEUP

F-6. The advisor must know HN sociopolitical and military organizations and their interrelationships, to include personalities, political movements, forces involved, and social drives. He must impress upon his counterpart the need for an integrated civil-military effort to defeat the insurgents. His counterpart must learn that military actions are subordinate to, and supportive of, the economic and social actions required to remove the insurgency's causes.

STATUS OF ADVISOR

F-7. The advisor must fully understand his status in the HN. Agreements between the United States and the HN spell out his status. These agreements may provide full diplomatic immunity or very little immunity. Without an agreement, the advisor is subject to local laws, customs, and the jurisdiction of local courts. Regardless of the diplomatic immunity afforded him, the advisor observes local laws, applicable laws of war, and Army regulations and directives.

RAPPORT

F-8. Rapport is a sympathetic relationship between people that is based on mutual trust, understanding, and respect. Personal dislike, animosity, and other forms of friction characterize the lack of rapport.

F-9. The need to establish rapport with HN counterparts is the result of a unique military position in which the advisor has no direct authority or control over their actions. However, an advisor can influence or motivate his counterpart to act in certain ways by using the proper advisory techniques.

F-10. Effective rapport must exist to gain the control needed to execute the mission. The successful advisor establishes rapport that allows influence over the counterpart's actions despite the absence of formal authority.

F-11. Rapport results when each individual perceives the other as competent, mature, responsible, and compatible (working toward a common goal). If the advisor can convey this attitude to his HN counterpart, long-lasting, effective rapport will exist.

TECHNIQUES

F-12. An advisor must always remember that he is an advisor and not a commander. He is not there to lead troops.

F-13. Having the counterpart select a particular COA is only possible if he perceives the advisor has the professional competence to give sound advice. If the counterpart does not believe the proposed solution to a problem is effective or realistic, he will question the advisor's competence. The advisor must explain to his counterpart why the advice is sound.

F-14. The advisor does not use bribery or coercion, since results achieved from these actions are only temporary. As soon as the "payment" is made, or the "force" is removed, the counterpart has no reason to comply. In practice, these techniques are not efficient and will not achieve the long-term goal of developing proficiency, competence, and initiative in the counterpart.

F-15. The advisor must be careful not to bribe or coerce a counterpart unintentionally. He must be aware that as an American Soldier he might have privileged status in the HN. The advisor's presence may garner personal benefits for the counterpart through his position of having a one-on-one association with an American. Conversely, the advisor may make a counterpart afraid of offending to the point of complying with every suggestion the advisor makes.

F-16. In short, psychologically pressuring a counterpart is not recommended. Such pressure is used only as a last resort, since it may irreparably damage the relationship between the advisor and his counterpart. However, psychologically pressuring the HN counterpart may sometimes be successful. Forms of psychological pressure may range from the obvious to the subtle. The advisor never applies direct threats, pressure, or intimidation on his counterpart. Indirect psychological pressure may be applied by taking an issue up the chain of command to a higher U.S. commander. The U.S. commander can then bring his counterpart to force the subordinate counterpart to comply. Psychological pressure may obtain quick results but have very negative side effects. The counterpart will feel alienated and possibly hostile if the advisor uses such techniques. Offers of payment in the form of valuables may cause him to become resentful of the obvious control being exerted over him.

F-17. Advising works both ways. The advisor sets an example for the counterpart by asking his advice. The advisor must realize that the counterpart is the expert in his country and that he can learn much from him.

F-18. The advisor must avoid giving the counterpart the impression that status reports and administrative requirements are the most important items. Such an impression may cause the counterpart to become aloof because it may be difficult and time-consuming for him to get this information. The advisor must treat his counterpart as an equal. He must also give the respect he himself expects to receive. He must take care not to make this fellow soldier feel like an errand boy.

F-19. The advisor transacts important business directly with his counterpart to ensure full understanding of difficult subjects. He uses the soft approach to request official information.

F-20. The advisor does not present too many subjects at one time or unnecessarily prolong the discussion of one subject. The advisor schedules another conference later, if needed.

F-21. The advisor corrects the most important deficiencies first. Upon his arrival in the AO, he will find many matters in need of immediate corrective action. He avoids telling his counterpart that everything is all wrong. Rather, he looks for the good systems and policies and praises his counterpart on his successes. At this point, the counterpart will normally point out deficiencies that need correction as his idea, and a joint problem-solving process can begin. In some cases, it may take a month or more to sell one idea.

F-22. When making recommendations, the advisor phrases them in a way that will not impose his will on the HN commander's decisions. The advisor leaves enough room for his counterpart to exercise his prerogative. One of his counterpart's greatest fears is that his troops will see him as dependent upon his advisor. The advisor carefully chooses a time and place to offer advice.

F-23. During combat operations, the advisor advises the commander but never usurps his command or authority. The amount of advising during combat operations is small. The advisor does most of his advising while preparing for combat. He bases his advice on his observations or those of his subordinates during past operations. He holds a private critique with the commander upon completion of an operation.

F-24. The advisor must not be afraid to advise against a bad decision. He does it tactfully, however. He acts as a staff member who recommends a change of action to an American commander he respects and with whom he works daily.

F-25. The advisor approaches the subject under discussion from different directions and with different words to make sure the advice given is clearly understood. He does not accept a "yes" answer at its face value. "Yes" may mean the person understands but does not necessarily accept the suggestion. It may also be used to cover a lack of understanding.

F-26. The advisor always exercises patience in dealing with a counterpart. He never expects a job to be done at the snap of a finger, and he does not snap a finger.

F-27. The advisor cannot accept information from his counterpart in blind faith. He checks it discreetly and diplomatically, but he must check it.

F-28. After the advisor plants an idea, he lets his counterpart take credit for it as if it were his own idea.

F-29. Advisors are transients. The advisor tries to learn what the previous advisor had tried and has or has not accomplished. He asks him for his files and thoroughly debriefs him to prevent reinventing the wheel. The advisor keeps an open mind and judges matters himself.

F-30. The advisor starts preparing a folder about the advisory area and duties as soon as possible. He maintains a worksheet-type folder during the tour to better understand the job. Follow-on advisors will have a complete file to assist them in completing projects.

F-31. The advisor does not hesitate to make on-the-spot corrections. He must be extremely tactful. Above all, he does not make the person he corrects lose face in front of his peers or subordinates. Embarrassing the counterpart, in most cultures, can cause a serious loss of rapport and possible mission failure. The advisor respects the almost universal custom and desire of "saving face."

F-32. An advisor must never make promises he cannot or must not carry out. He never pledges U.S. assets unless he has the authority and capability to deliver them.

F-33. Once advisors are committed, their activities should be exploited. Their successful integration into the HN society, their respect for local customs and mores, and their involvement with CA projects are constantly brought to light. In formulating a realistic policy for the use of advisors, the commander must carefully gauge the psychological climate of the HN and the United States.

PERSONAL QUALITIES

F-34. Advisors must rely on their abilities to sell the most indefinite commodity—themselves. The traits of an advisor encompass all the traits of leadership plus the ability to adapt to his environment. This environment changes with the assignment area. To sell himself, the advisor must prove his value and present a favorable personality in the eyes of his counterpart. This selling occurs in time by gradually demonstrating his capabilities in an unassuming but firm manner.

F-35. The advisor avoids rushing personal acceptance by the counterpart. Overselling himself will arouse suspicion and delay acceptance. Time spent developing a healthy relationship will pay large dividends later.

F-36. An advisor must be extremely flexible, patient, and willing to admit mistakes. He must persevere in providing sound advice. He must also be a diplomat of the highest caliber and possess an unusual amount of tact. An advisor must be honest. He must maintain high moral standards and be understanding and sincere. He must present a good military appearance, stay in good physical condition, and lead by example.

F-37. The advisor must know thoroughly the organization, equipment, and tactics of the unit he advises. He must be professional and proficient. He must demonstrate an awareness of his counterpart's problems.

F-38. The advisor must be positive, but not dogmatic, in his approach to any subject. If, however, he is not sure of the subject matter, he says so and takes the steps to obtain the correct information. He does not try to bluff his way through a problem.

F-39. Persistence, balanced with patience, is a favorable trait of an advisor. If he discovers a problem, he tries to solve it; he recommends the proper measures to take and then follows through. Patience is of utmost importance. He continually brings the matter to his counterpart's attention until he sells him on taking the measures to solve the problems or correct the deficiency. Ultimately, the goal is to advise his counterpart in such a way that he takes the desired action feeling that it was through his own initiative rather than the advisor's.

F-40. A successful advisor must have subject knowledge, the ability to demonstrate his capabilities in an unassuming but convincing manner, and the clear indication of his desire to get along with counterparts and other associates. Common sense is possibly the greatest asset of the successful advisor. Ultimately, this uncommon commodity separates the effective advisor from the ineffective one. With common sense, everything is possible; without it, failure can be expected.

ADVISORY GOALS

F-41. The advisor emphasizes in-place training when the units return to garrison (focus on battle drills and SOPs). Twenty-five-meter firing ranges are ideal to conduct marksmanship training (zero, reduced range qualification, night firing, and instinctive firing techniques).

F-42. The advisor spends maximum time with the unit so the troops get to know and trust him. He talks to and gets to know the troops, not just the unit leaders. He gets excellent feedback in the common soldier's candid comments. Such comments often reflect troop morale and operational effectiveness. He stays abreast of what is going on in the unit. He also stays in close contact with the commander and staff.

F-43. The advisor encourages frequent command inspections by the commander. In some HNs, this action is a new concept or an uncommon practice. Many HN commanders are reluctant to inspect. They rely solely on correspondence and reports to evaluate unit effectiveness.

F-44. The advisor continually stresses the obvious advantages of good military-civilian relations to avoid the idea of military arrogance, which easily irritates the civilian populace. The development of a proper soldier-civilian relationship is a critical factor in IDAD and in COIN. Improper behavior by soldiers toward civilians must be immediately corrected.

F-45. The advisor constantly strives to raise the HN units' standards to the level needed to complete the mission. He guards against lowering his standards but realizes most HN units needing advice may not have the logistic, educational, or nutritional base to perform to U.S. standards and, in that sense, may not be expected to meet U.S. standards.

F-46. The advisor keeps training standards high enough so that the unit is prepared for combat at all times. He does not use training time for housekeeping matters.

F-47. The advisor stresses human rights and the consequences of mistreating suspects and prisoners. The advisor constantly promotes unit esprit de corps to sustain the unit in the face of difficulties. The advisor persuades the HN personnel to pass information up, down, and laterally.

PERSONAL ATTITUDES AND RELATIONS

F-48. Becoming accustomed to the native food and drink, in somewhat varying degrees, poses a problem to the advisor. An advisor establishes and maintains rapport more easily by drinking in moderation and eating with counterparts IAW culturally acceptable rules. Refusal to accept their drink and food is often considered an insult.

F-49. The advisor does not become discouraged. All advice will not be accepted. Some will be implemented later.

F-50. The advisor cannot forget that a careless word or action can cost the United States dearly in good will and cooperation that may have been established with great effort and at considerable cost. The advisor does not criticize HN policy in front of HN personnel. It is the advisor's obligation to support the incumbent government just as he does his own. This obligation is U.S. national policy.

F-51. The advisor studies his counterpart to determine his personality and background. He makes every effort to establish and maintain friendly relationships. He learns something about his counterpart's personal life and demonstrates an interest in his likes and dislikes.

F-52. He sets a good example in dress, posture, and personal conduct and in professional knowledge and competence. He emphasizes the importance of doing things on time by demonstrating punctuality. Many cultures have a very casual attitude toward time. He realizes, however, that he will never change their culture but may succeed in modifying their behavior to meet mutually recognized mission needs.

F-53. He develops a sense of responsibility toward the unit he advises to the degree that he senses personal fulfillment for a job well done. He avoids the pitfall of becoming so involved with the unit that he cannot readily recognize failures.

F-54. The advisor accepts invitations to dinners, cocktail parties, and ceremonies. He engages in cordial social conversation before discussing business matters. He only discusses business matters when appropriate.

F-55. The advisor recognizes and observes military courtesy and local customs and courtesies. When in doubt, he leans toward the polite. The advisor does not get caught in personality clashes between HN officers who may concern themselves more with person-to-person relationships than with organizational frameworks.

ADVISOR CONSIDERATIONS

F-56. Following is a list of suggestions and considerations that will benefit advisors. Advisors participate in MCA programs and tactical, intelligence, and PRC operations.

MILITARY CIVIC ACTION PROGRAMS

F-57. In MCA programs, the advisor considers the following:

- *Communications.* The advisor must get his ideas and intentions across through his counterpart. Programs can be publicized by—
    - Community meetings.
    - News media.
    - Informal lectures.
    - Demonstrations.
- *Image.* In many areas, relations between villagers and the government may not always have been satisfactory. The government should—
    - Establish rapport with the people.
    - Speak their dialect.
    - Understand their culture.
    - Be sympathetic to their problems.
- *Demonstrations.* The government shows the villagers how a dynamic program works. The populace is encouraged to participate voluntarily in projects to—
    - Instill a feeling of ownership and responsibility.
    - Teach the populace how to maintain them.
- *Traditions.* Projects are based on local traditions and customs so that the populace does not become skeptical of them.
- *Timeliness.* Major work projects are started and completed during seasonal unemployment, not during planting or harvesting time.
- *Flexibility.* Projects are altered if unforeseen conditions arise.
- *Continuity.* The government must instill in the populace confidence that it intends to see the project through.
- *Maintenance.* The people must be left with the means and know-how to maintain the project. Repair parts must be available after the government representatives depart. Procuring manufactured materials and expertise locally ensures the maintenance of the project.

TACTICAL OPERATIONS

F-58. In tactical operations, the advisor considers the following:

- Orient on the threat, not on the terrain.
- Maintain the offensive, regardless of the weather.
- Establish priorities of effort.
- Operate in the threat environment.
- Enforce the concept of subordinate units backbriefing their plan to higher HQ.
- Emphasize secrecy and surprise. Plans should provide for—
    - Effective and secure communications.
    - Constant indoctrination of the individual soldier.
    - Variation of TTP to avoid establishing patterns.
- Emphasize command and staff actions that include—
    - Centralized planning of small-scale, decentralized tactical operations.
    - Integrated planning, to include MCA, PSYOP, and PRC operations. (If possible, civil defense or local law enforcement agencies, not the military, conduct PRC operations.)

- Ensuring unity of command.
- Ensuring training programs are designed to develop the offensive spirit, physical stamina, and desire to seek out and destroy the threat and to train paramilitary forces for security operations.
- Planning for the use of reserve forces.
- Planning and executing the intelligence collecting effort by coordinating the integration of all available agencies and interrogating prisoners and suspects.
- Providing for the rapid collection and dissemination of all available information and intelligence so that forces can take immediate action to destroy a fast-moving threat.
- Integrating detailed logistics into all tactical planning.
- Judiciously applying firepower in view of the minimum destruction concept to reduce the alienation of the populace.
- Using all means of mobility, to include aircraft, tracked and wheeled vehicles, boats, animals, and porters.
- Ensuring communications requirements are based on the HN capabilities—requirements for amplitude modulation (AM), frequency modulation (FM), and single sideband (SSB); air-to-ground (FM, ultrahigh frequency [UHF], very high frequency [VHF], or SSB) for C2, close air support (CAS), radio relay, and MEDEVAC; fire support plans; and emergency nets in various regions.
- Ensuring the adequate support of attached, nonorganic forces.

INTELLIGENCE OPERATIONS

F-59. The advisor evaluates—

- The S-2 and/or intelligence section and its operating procedures and effectiveness.
- The personalities, counterparts, and other persons with whom business is conducted.
- The chain of command and communications channels of the HN.
- The intelligence projects begun by predecessors.
- The intelligence projects predecessors believed should have been initiated.
- The advisor communications channels.
- The reference material available.
- The other intelligence agencies.
- The intelligence collection by enforcing the concept of what CCIR are and how each part of the force supports these commander's priorities.

Note. The advisor prepares and maintains a list of PIRs and/or IRs and threat indicators.

POPULACE AND RESOURCES CONTROL OPERATIONS

F-60. Advisors help their counterparts develop proper control plans and training programs for PRC measures. Advisors also help coordinate plans and requests for materiel and submit recommendations to improve the overall effectiveness of operations. Advisors can be helpful in—

- Preparing to initiate control. They—
  - Select, organize, and train paramilitary and irregular forces.
  - Develop PSYOP activities to support PRC operations.
  - Coordinate activities through an area coordination center (if established).
  - Establish and refine PRC operations.

- Intensify intelligence activities.
- Establish and refine coordination and communications with other agencies.
- Establishing maximum control. Continued threat success will dictate the intensification of control measures. Advisors—
  - Establish defended villages (civil defense sites) and relocate populace (as a last resort).
  - Initiate and publicize amnesty and rehabilitation programs.
  - Offer rewards for the capture and defection of insurgent cadres.
  - Establish martial law.
- Relinquishing control. As internal defense succeeds, controls are reduced in two stages. In Stage A, advisors reduce the intensity of controls by—
  - Continuing general area controls but reducing raids, ambushes, and cordon and search.
  - Passing primary responsibility for control to police and paramilitary units, phasing out military participation.
  - Continuing intelligence activities.
  - Accelerating internal development.
  - Taking maximum psychological advantage of reduced control.
- In Stage B, advisors reduce control activities to a minimum by—
  - Lessening individual restrictions.
  - Continuing controls on resources and populace movements.
  - Continuing intelligence and PSYOP programs.
  - Emphasizing internal development and political allegiance.
- Making provisions for handling, accounting for, and disposing of insurgents, sympathizers, suspects, and other violators and confiscated contraband. These provisions include—
  - Setting up detention and interrogation facilities.
  - Recording the circumstances of capture to analyze trends and patterns.
  - Handling prisoners referred for prosecution or rehabilitation.
  - Documenting, safeguarding, and turning over confiscated materiel to the proper authorities.
- Establishing amnesty, pardon, rehabilitation, reward, and reeducation programs. Reward programs are begun and payments provided for information leading to the capture of insurgents, weapons, and equipment. Amnesty and rehabilitation programs must include the following:
  - Provisions to allow individuals to again support the government without fear of punishment for previous antigovernment acts, wherever possible.
  - Just and equitable programs to induce disaffection among insurgents and their supporters.
  - Rehabilitation of former insurgents and their supporters through reeducation and constructive, controlled employment.

Intelligence Operations

Intelligence is an integral part of FID operations. Intelligence is a four-phase, cyclical process. Order of battle (OB) intelligence occurs in two phases and requires more detailed intelligence at the lower echelons.

ROLE OF INTELLIGENCE

G-1. The primary duty of intelligence personnel engaged in FID is to produce intelligence to prevent or defeat lawlessness or insurgency. The SF unit must be ready to train, advise, and assist HN personnel in intelligence operations. Intelligence personnel must collect information and produce intelligence on almost all aspects of the FID environment. When they know that insurgents, terrorists, or common criminals receive aid from an external power, intelligence personnel seek information on the external power's role in the insurgency. They need information not only on the armed insurgents but also on their infrastructure organizations and their relationships with the populace. These relationships make the populace a most lucrative source of information.

G-2. A sound collection program and proper use of the various collection agencies and information sources will result in a very heavy volume of information flowing into the intelligence production element. Because of the insurgent environment, politics, and military tactics, intelligence personnel can meet intelligence requirements only by reporting minute details on a great variety of subject areas. Each detail may appear unrelated to others and insignificant by itself. However, these details, when mapped and chronologically recorded over long periods and analyzed with other reported details, may lead to definitive and predictable patterns of insurgent activity.

G-3. The insurgent recognizes the shortcomings in his military posture. Therefore, he must minimize the weaknesses inherent in using and supporting isolated, unsophisticated forces that use ponderous and primitive communications and logistics systems. He uses the weather, terrain, and populace, employing secrecy, surprise, and simplicity. Plans and actions these unsophisticated forces will carry out must be simple, comprehensive, and repetitive. Therefore, the solution to a problem is a system that as a whole is complex but in part is independent, having simple, logical, and uniform characteristics.

INTELLIGENCE REQUIREMENTS

G-4. Accurate, detailed, and timely intelligence is vital to successful FID operations. This dependence on intelligence and CI is greater in FID operations than in conventional operations because of the differences addressed below.

FOREIGN INTERNAL DEFENSE OPERATIONS

G-5. In FID operations, the targets are elements of the populace—either civilian supporters or members of the insurgency. The differences between supporters and members are usually ill-defined. A complete awareness and intimate knowledge of the environment is essential to conducting current intelligence operations. The basic nature of the internal security problem requires an intensive initial intelligence effort to pinpoint the roots of subversion.

CONVENTIONAL OPERATIONS

G-6. In conventional operations, a force may succeed in capturing a military objective by attacking with overwhelming strength. A force can sometimes attain success in these situations without timely and detailed intelligence. Such success is not the case in FID. The insurgents seldom hold terrain. They will not overtly commit themselves except when cornered or when the odds heavily favor their chances of winning. Most importantly, their base of operations is in the populace itself. The insurgents, therefore, cannot be easily detected and overwhelmed. Insurgents require close scrutiny, delicate and discriminating analysis, and aggressive and accurate countermeasures.

G-7. The intelligence required is of the type, quantity, and quality that—

- Provides goals for daily or major operations (intelligence that locates guerrillas for tactical counterguerrilla operations).
- Enables HN forces to retain or regain the initiative.
- Enables HN forces to put continuous and increasing pressure on insurgent security.

THE INTELLIGENCE CYCLE

G-8. Intelligence operations follow a continuous, four-phase process known as the intelligence cycle (Figure G-1). The intelligence cycle is oriented to the commander's mission. Supervising and planning are inherent in all phases of the cycle. Even though the four phases take place in sequence, intelligence analysts perform all concurrently. While intelligence analysts process available information, the intelligence staff collects additional information, planning and directing the collection effort to meet new demands. The intelligence staff disseminates the intelligence as soon as it is available or needed.

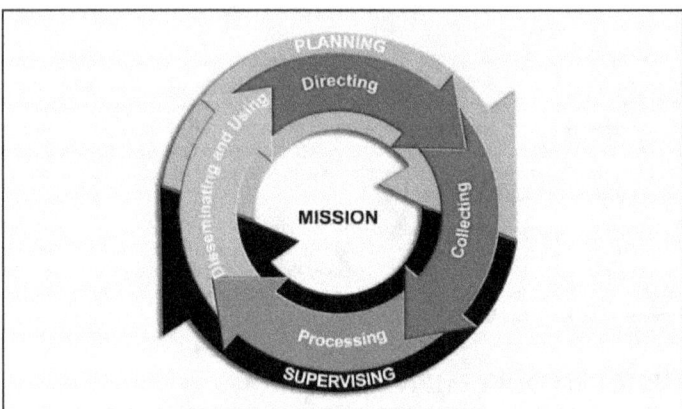

Figure G-1. The intelligence cycle

DIRECTING

G-9. The HN commander, through his senior intelligence officer (SIO), directs the intelligence effort. The SIO supervises collection management before the operation and guides the effective use of collection assets during the operation. He develops and maintains databases through research and intelligence preparation of the battlefield (IPB). IPB, coupled with the available database, provides a basis for situation and target development. The IPB and database provide a way to project battlefield events and activities in the AO and to predict COAs. By comparing these projections with actual events and activities as they occur, the SIO can provide the commander with timely, complete, and accurate intelligence.

G-10. Intelligence agencies, from national level down, constantly develop and maintain intelligence databases. The SIO accesses these databases to prepare initial intelligence estimates and to analyze the AO showing probable COAs. The SIO bases this analysis on the mission requirements and the commander's

PIRs. Intelligence analysts integrate the resulting intelligence estimate with other staff estimates and present them to the commander. The commander can then decide what actions he must take to perform the mission.

COLLECTING

G-11. The commander and his S-2 and S-3 begin the collection effort by determining requirements and establishing their priorities. They may base their requirements on mission, enemy, terrain and weather, troops and support available—time available (METT-T) and the commander's planning guidance.

G-12. PIRs are the basis for intelligence operations. The commander personally approves them. PIRs are those intelligence requirements for which a commander has an anticipated and stated priority in his task planning and decision making. In essence, the SIO organizes his PIRs and IRs as follows:

- He subdivides strategic PIRs and IRs into military, political, economic, psychological, and social categories, focusing on the national or international level.
- He subdivides operational PIRs and IRs into military, psychological, and social categories, focusing on the provincial or subnational level.
- He subdivides tactical PIRs and IRs into military, psychological, and social categories, focusing on the local level.

G-13. Within each of the above categories, the SIO identifies the specific discipline or disciplines that can be best used to collect needed information. The disciplines include—

- Human intelligence (HUMINT).
- Signals intelligence (SIGINT).
- Imagery intelligence (IMINT).
- Technical intelligence (TECHINT).
- Measurement and signature intelligence (MASINT).
- Open-source intelligence (OSINT).

Note. Of these disciplines, HUMINT, SIGINT, and IMINT are usually of greatest use to SF.

G-14. Specific information requirements (SIRs) are the specific items of information needed to satisfy PIRs and IRs. They are the basis for collection operations.

G-15. Intelligence analysts advise the SIO on the PIRs and IRs. They analyze METT-T and the commander's guidance and concept of the operation to determine needed information or intelligence. They review the existing database to identify available information and information to be acquired. They pass requirements for new information to the collection management and dissemination (CM&D) section as collection requirements.

G-16. Based on requirements, the CM&D section manages the collection effort. The section develops a collection plan keyed to the METT-T, the commander's concept of the operation, and the current situation. The section continuously updates the collection plan.

G-17. In FID operations, the problem is to identify and then locate the enemy. As frequently stated, in an insurgency the front is everywhere. Even after identifying and establishing operation patterns of members of the underground, the local police or security force must locate the enemy before they can capture them. There are essentially three methods of obtaining contact intelligence:

- *Patrols.* After developing some knowledge of the behavioral patterns of the underground or guerrillas from a study of their past movements, patrols or police squads can search for physical evidence (tracks and campsites). If there is a consistent pattern, patrols can be selectively dispatched based on anticipated movement of the insurgents.
- *Forced contacts.* When the guerrillas are separated from the people, their normal underground supply channels are cut off. This separation forces the guerrillas into the open to contact their underground and auxiliary elements. After identifying members of the underground and the

auxiliary, the police can arrest them. The guerrillas will then have to visit the remaining members of the underground and auxiliary more often to get required support.

- *Informants.* Using informants is a reliable and quick means of obtaining specific data required in contact intelligence. Through a process designed to protect their identity, informants pass information about movements, positions, and activities of the insurgents almost immediately. The local security force receives this information. Its commander should be authorized to take immediate action on his own authority with no requirement to seek approval from higher authorities.

G-18. Intelligence personnel must consider the parameters within which a revolutionary movement operates. Frequently, they establish a centralized intelligence-processing center to collect and coordinate the amount of information required to make long-range intelligence estimates. Long-range intelligence focuses on the stable factors existing in an insurgency. For example, various demographic factors (ethnic, racial, social, economic, religious, and political characteristics of the area in which the underground movement takes place) are useful in identifying the members of the underground. Information about the underground organization at national, district, and local levels is basic in FID operations.

G-19. Collection of specific short-range intelligence about the rapidly changing variables of a local situation is critical. Intelligence personnel must gather information on members of the underground, their movements, and their methods. Biographies and photos of suspected underground members, detailed information on their homes, families, education, work history, and associates are important features of short-range intelligence.

G-20. Destroying its tactical units is not enough to defeat the enemy. Forces must neutralize the insurgent's underground cells or infrastructure first because the infrastructure is his main source of tactical intelligence and political control. Eliminating the infrastructure within an area achieves two goals: it ensures government control of the area and cuts off the enemy's main source of intelligence. An intelligence and operations command center is needed at district or province level. This organization becomes the nerve center for operations against the insurgent infrastructure. Information on insurgent infrastructure targets should come from such sources as the national police and other established intelligence nets, agents, and individuals (informants).

G-21. Security forces can induce individuals among the general populace to become informants. Security forces use various motives (civic-mindedness, patriotism, fear, punishment avoidance, gratitude, revenge or jealousy, and financial rewards) as persuasive arguments. They use the assurance of protection from reprisal as a major inducement. Security forces must maintain the informant's anonymity and must conceal the transfer of information from the source to the security agent. The security agent and the informant may prearrange signals to coincide with everyday behavior.

G-22. Surveillance, the covert observation of persons and places, is a principal method of gaining and confirming intelligence information. Surveillance techniques naturally vary with the requirements of different situations. The basic procedures include mechanical observation (wiretaps or concealed microphones), observation from fixed locations, and physical surveillance of subjects.

G-23. Whenever a suspect is apprehended during an operation, a hasty interrogation/tactical questioning takes place to gain immediate information that could be of tactical value. The most frequently used method for gathering information (map studies and aerial observation), however, is normally unsuccessful. Most PWs cannot read a map. When PWs are taken on a visual reconnaissance flight, it is usually their first flight and they cannot associate an aerial view with what they saw on the ground.

G-24. The most successful interrogation method consists of a map study that is based on terrain information received from the detainee. The interrogator first asks the detainee what the sun's direction was when he left the base camp. From this information, he can determine a general direction. The interrogator then asks the detainee how long it took him to walk to the point of his capture. Judging the terrain and the detainee's health, the interrogator can determine a general radius in which the base camp can be found (he can use an overlay for this purpose). He then asks the detainee to identify significant terrain features he saw on each day of his journey (rivers, open areas, hills, rice paddies, and swamps). As the detainee speaks and his memory is jogged, the interrogator finds these terrain features on a current map and gradually plots the detainee's route to finally locate the base camp.

G-25. If the interrogator is unable to speak the detainee's language, he interrogates through an interpreter who received a briefing beforehand. A recorder may also assist him. If the interrogator is not familiar with the area, personnel who are familiar with the area brief him before the interrogation and then join the interrogation team. The recorder allows the interrogator a more free-flowing interrogation. The recorder also lets a knowledgeable interpreter elaborate on points the detainee has mentioned without the interrogator interrupting the continuity established during a given sequence. The interpreter can also question certain inaccuracies, keeping pressure on the subject. The interpreter and the interrogator must be well trained to work as a team. The interpreter has to be familiar with the interrogation procedures. His preinterrogation briefings must include information on the detainee's health, the circumstances resulting in his detention, and the specific information required. A successful interrogation is contingent upon continuity and a well-trained interpreter. A tape recorder (or a recorder taking notes) enhances continuity by freeing the interrogator from time-consuming administrative tasks.

PROCESSING

G-26. Processing is the step in the intelligence cycle through which information becomes intelligence. It consists of recording, evaluating, integrating, and interpreting. Certain factors are unique to the internal defense environment. Intelligence analysts must apply these factors to determine insurgent capabilities and COAs and provide the intelligence needed for all facets of FID operations. An often-overlooked technique of determining what the insurgents are doing to influence the population can be found in open-source media. Analysts should make a concerted effort to review and capture any and all insurgent messages found on the Internet, television, radio broadcast, and newspapers, as well as signs placed on walls of buildings in built-up areas. Once collected, these messages should be tracked and measured against future actions of the insurgents and population. These messages support understanding of the insurgents' thought process that can be considered in planning counterinsurgent action.

Recording

G-27. Like conventional tactical situations, FID operations require large amounts of information on a continuous basis. Intelligence analysts promptly compare this information with existing information and intelligence to determine its significance. To a large degree, the extent of the recording effort will depend upon the insurgent activity in the area.

G-28. Depending on the echelon of responsibility, the state of insurgent activity in the area, and the degree of knowledge of the enemy, the current intelligence graphic requires at least two annotated maps: the incident map and the insurgent situation map (SITMAP). Each of these recording devices normally is a transparent overlay covering a large-scale topographic map of the area. The incident map provides historic, cumulative information on insurgent activity trends or patterns. Properly maintained entries let the intelligence analyst make judgments about—

- The nature and location of insurgent targets.
- The relative intensity of insurgent interest in specific areas.
- Insurgent control over or support from the population.
- Potential areas of insurgent operations.

G-29. The insurgent SITMAP represents intelligence; much of the SITMAP is built around the information recorded on the incident map. Intelligence analysts will find it difficult to pinpoint insurgent installations and dispositions with the same degree of confidence as in a conventional tactical situation. The insurgents can displace on short notice, making a report outdated before it can be confirmed. The SITMAP can graphically substantiate the trends or patterns derived from the incident map, which improves the economy and effectiveness of the collection effort. The SITMAP provides a ready guide for briefing the commander, the civil authorities, or other interested parties.

G-30. Other annotated maps include the trap map and personalities and contact maps. The trap map is used if the insurgent is capable of sabotage or terrorist action. It will portray particularly attractive target locations for insurgent sabotage or terrorism. Insurgent targets could be road and railroad bridges, communications centers, theaters and assembly halls, and places where the terrain favors ambushes and raids. These areas are plainly marked on this map, directing attention to possible insurgent access and escape routes.

G-31. Initial intelligence about the insurgent situation may be information on locations and activities of individual agents (espionage, agitation, organization, and liaison). The personalities and contacts map records the appearances, movements, meetings, and disappearances of these agents. A large-scale city street map or town plan is required to track the individuals. Dated symbols indicate observations and incidents. Depending on the amount of insurgent activity, intelligence analysts can combine this map with the incident map.

G-32. The intelligence worksheet and the annotated maps serve to isolate problem areas and form ties between items of information and intelligence collected. In the early phase of an insurgency, the enemy is building his own organization. His organizational procedures and tactics will, therefore, be unique. The intelligence analyst must study personalities and analyze incidents.

G-33. The insurgency analysis worksheet helps identify information and intelligence needed to satisfy PIRs and IRs. It provides a guide for analysis of an environment for operations short of war.

G-34. The hot file is the most important working file. It includes all available material pertaining to an incident or groups of possibly related incidents of current interest. This file contains material on persons, agents or suspects, or places likely to be involved in insurgency activity.

G-35. If propaganda is a major part of the insurgent effort in the area, a current propaganda and PSYOP file should contain items pertaining to the grievances insurgents are exploiting, such as—

- Literature.
- Background material.
- Propaganda speeches.
- Analyses of local grievances.

G-36. Each insurgent personality has a local personality and organization file. If the local police force carries out surveillance, they can transfer basic identifying and biographical information from dossiers to a card file. This card file helps train friendly surveillants to recognize key personalities on sight. The organization section of this file contains information on—

- The history and activities of the fronts for the insurgent organization.
- Other subversive or suspected groups and their officers.
- Overlapping directorates of, membership in, and liaison among insurgent fronts.

G-37. The area study files contain up-to-date and pertinent data on the—

- Geography.
- Hydrography.
- Climate.
- Political and economic characteristics.
- Civil populace.
- Military and paramilitary forces.
- Resistance organization.
- Targets.
- Effects of the above-listed characteristics.

G-38. A resource file contains all material of importance but not of immediate value. It may include inactive incident files, inactive personality and organization files, and photography.

Evaluation

G-39. Evaluation is the examination of information to determine its intelligence value. The intelligence analyst's knowledge and judgment play a major role in evaluating information. Therefore, he must know the theory of insurgency. In considering if a fact or event is at all possible, he must realize that certain events are possible, although they have not previously occurred and have been thought unlikely to occur. Initially, intelligence production starts with unconfirmed information that is subsequently supported, confirmed, or denied by additional and related information. As the intelligence analyst obtains more information, the insurgent situation, capabilities, and probable COAs become increasingly clear.

Integration

G-40. Evaluated information becomes intelligence only after intelligence analysts have integrated it with other information and interpreted it to determine its significance. Integration involves combining selected data to form a pattern that will have meaning and establish a basis for interpretation. In his search for related information, the intelligence analyst checks the incident file, the friendly and suspect personality files, and the organizational file. After obtaining all related items of information from the intelligence files, he begins to assemble the available information to form as many logical pictures or hypotheses as possible. Alternative methods of assembly are an essential prerequisite to any valid interpretation. The assembly of information to develop logical hypotheses requires good judgment and considerable background knowledge. In formulating hypotheses, the intelligence analyst must avoid limitations resulting from preconceived opinions.

G-41. The intelligence analyst uses the IPB process for intelligence production. The IPB supports commanders and their staffs in the decision-making process. The commander directs the IPB effort through the CCIR (which, for the SIO and his intelligence analysts, include the PIRs and IRs). All other staff elements are active participants in the IPB. FM 3-05.102, *Army Special Operations Forces Intelligence*, and FM 34-130, *Intelligence Preparation of the Battlefield*, contain detailed discussions of the IPB process.

Interpretation

G-42. Interpretation is the result of deducing the probable meaning of new information and determining its implications about future insurgent activities. The meaning of the information is determined in relation to the insurgent situation and the insurgents' probable COAs.

DISSEMINATING AND USING

G-43. The final step of the intelligence cycle is disseminating and using the intelligence processed. Intelligence and combat information are of little value if not delivered when needed. Failure to disseminate this intelligence defeats a thorough and successful collection and processing effort. Because of IPB, the SIO produces a variety of templates, overlays, association and event matrixes, and flowcharts appropriate to METT-TC. He provides these products to the HN commander and S-3 for approval and guidance. As a follow-up, the SIO provides the correct products promptly to the right consumers. He also ensures these products are adequate for and properly used by them. Where appropriate, the SIO must advise and coach nonintelligence personnel in their use. He must also use his IPB products to identify gaps in the intelligence database and redirect his collection effort.

THREAT ANALYSIS

G-44. Threat analysis focuses on the examination of the insurgents' ends, ways, means, vulnerabilities, centers of gravity, and friendly methods for gaining the initiative, exploiting success, and achieving early victory. Insurgents are potentially quite vulnerable in some areas. The insurgents—

- Are normally outnumbered and outgunned by the security forces, although they may have local fire superiority.
- Are deficient in mobility, communications, medical, and logistical support.

- Are considered illegal by the government in power.
- Lack a stable political, economic, and territorial base.

G-45. Insurgents are aware of their difficult situation; therefore, they must protect and overcome their vulnerabilities. They must maintain security while building strength and support. They can do this by—

- Developing underground organizations and support systems.
- Infiltrating government organizations for intelligence and political purposes.
- Gaining the willing or unwilling support of the populace for intelligence, logistics, and manpower.
- Establishing remote base areas.
- Using multiple secret routes.
- Using mines and booby traps.
- Using caches.

G-46. The insurgents must gain and maintain the initiative by carrying out actions that distract security forces (forcing the security forces to take a defensive posture). They also can gain and maintain the initiative by carrying out actions that weaken the government in power. They weaken the government by attacking its political and economic infrastructure through acts of terror, military attacks against economic targets, and the skillful use of propaganda.

G-47. Security is essential for the insurgents' success, because it provides them with the time to make a long-term strategy work. To do so, they must protect their vulnerabilities and weaknesses and maintain the ability to exercise the initiative. Security is the insurgents' true center of gravity. The government must use intelligence to expose vulnerabilities, regain the initiative, and destroy the insurgency-developed and intelligence-oriented strategy. The HN forces must focus their efforts on planning and conducting operations that reduce the insurgents' freedom of action and attack the insurgents' vulnerabilities.

ORDER OF BATTLE INTELLIGENCE

G-48. OB is as important in an insurgency as in conventional combat operations. However, the intelligence analyst must recognize some differences in nomenclature and approach. The applicability of the OB factors differs in an insurgency from conventional operations. There will also be differences in application between Phase I and Phase II insurgency situations. The elements of the OB factors are dependent on each other. They are closely related and must be considered as a whole. Information on one of the elements will often lead to a reevaluation or alteration of information previously received on another element. The normal practice of developing and maintaining OB down to and including two echelons below the intelligence analyst's own level of command does not apply to FID. The nature of the insurgency and the phased development of its forces require much more detailed OB and pertain to much lower echelons. The following paragraphs address the OB factors and explain their applicability to insurgency situations.

COMPOSITION

G-49. In some insurgent movements, military force is only one of several instruments through which the insurgents seek power. Development of a military force has the lowest priority during the early stages of an insurgency. As long as the party core and civil organizations are established and move effectively toward the goal of the insurgency, the military arm may either be dormant or simply exist in cadre form until needed as a support arm.

Phase I Considerations

G-50. Rather than collecting information on the identification and organization of specific insurgent units, the intelligence personnel concentrate on the internal workings of insurgent activity groups. Knowledge of their composition can be a key to the entire planned course of the insurgency. Details of composition may include the appearance of new organizations, the relative amount of enemy effort in rural and urban operations, the internal C2 chain, and the organization of the insurgent groups.

Phase II Considerations

G-51. The concern of the intelligence analyst will be to determine the composition of the insurgent combat units (including their organization and C2). The degree of sophistication encountered indicates other factors (training, logistics, and strength). Armed platoons or small terrorist cells indicate the overt military portion of the insurgency plan is just beginning. Armed battalions and large urban terrorist groups indicate there is a serious menace to the current government.

POLITICAL STRUCTURES

G-52. A tightly disciplined party organization, formally structured to parallel the existing government hierarchy, may be found at the center of some insurgent movements. In most instances, this organizational structure will consist of committed organizations at the village, district, province, and national levels. Within major divisions and sections of an insurgent military HQ, totally distinct but parallel command channels exist. There are military chains of command and political channels of control. The party ensures complete domination over the military structure by using its own parallel organization. The party dominates through a political division in an insurgent military HQ, a party cell or group in an insurgent military unit, or a political military officer.

COMBAT FORCES

G-53. The organization of insurgent combat forces is dependent on the needs, the tactics used, and the availability of personnel and equipment. Frequently, subordinate elements of insurgent units employ independently. The intelligence analyst who receives a confirmed report of an insurgent unit operating in his area cannot, therefore, assume that the parent unit is also present.

DISPOSITION

G-54. Determining the disposition of the insurgents involves locating their operational training and supply bases, LOCs, and areas of political control. The intelligence analyst can arrive at the insurgents' potential dispositions by developing patterns of activity. These patterns originate from map study and knowledge of insurgent tactics. Insurgent base areas, for instance, are normally near areas the insurgents control politically, thereby providing an early warning system. By plotting insurgent sightings and combining this information with weather conditions, time factors, detailed investigation of insurgent incidents, and AARs, the intelligence analyst can select possible enemy dispositions and possible areas of tactical deployment. These areas, while appearing to be under the control of internal defense forces, may be under the political control of the insurgents.

Phase I Considerations

G-55. This phase considers the location, deployment, and movements of insurgent organizations or personnel. The insurgents' strength and tactics may be revealed, to some extent, by discovering whether they concentrate their effort in a few places or disperse throughout the target nation. If they initially concentrated their effort in one city or in a rural area, then the spread of the insurgent organization is a key to how long they have been operational and how successful they have been.

Phase II Considerations

G-56. How the insurgent forces are deployed can indicate whether the enemy is making a widespread show of strength (with units scattered about the country) or is concentrating his forces around a few key targets. The deployment can also show whether the enemy is going to concentrate on such activities as interdicting transportation or actively seeking battle with government forces.

STRENGTH

G-57. Intelligence analysts must think of the strength of the insurgent forces in terms of the combat forces, political cadres, and popular support. The intelligence analyst can apply conventional methods of strength computation to determine insurgent strength. The insurgents will try to have their strengths overestimated by the HN security elements. To give this false impression, the insurgents will rapidly move their units and use multiple designations for a single element. The intelligence analyst views reports from the populace on insurgent strengths with caution and stresses the importance of actual counts of enemy personnel. He finds it more difficult to determine the popular support for the insurgents, although a guide may be the percentage of an area under government control as opposed to the percentage under insurgent control. A useful indicator of the extent of insurgent political control is the willingness of the populace to report information on the insurgents.

Phase I Considerations

G-58. The cadre that organizes and activates the movement usually consists of highly trained, aggressive professionals who exercise an influence out of proportion to their actual numbers. The intelligence analyst's concern is with the number of units in existence. In this phase, the intelligence analyst identifies and evaluates new groups and organizations that have appeared in the nation and the changes in the size of existing groups.

Phase II Considerations

G-59. The actual number of men available to the insurgency now assumes the importance it lacked, to some degree, in Phase I. By knowing the amount of weapons and equipment the insurgents have, the intelligence analyst can estimate their capabilities against friendly forces. The degree of popular support for the insurgents will be manifested in such areas as recruiting for their forces.

TACTICS

G-60. Tactics include enemy doctrine and the conduct of operations according to that doctrine. Insurgent forces may be more flexible in their application of doctrine than regular military organizations. The friendly forces must know and understand the doctrine that guides the insurgents if they are to counter enemy efforts effectively. The choice and application of insurgent tactics is an appraisal of friendly and insurgent strengths. Insurgent tactics will involve political, military, psychological, and economic considerations, all closely integrated. Speed, surprise, and heavy application of firepower and mobility describe military tactics.

Phase I Considerations

G-61. An absence of strictly military operations and an emphasis on subversion and organizational development describe this phase. Although instances of terrorism may begin to occur in the latter stages of Phase I, military activity is usually limited to recruiting and establishment of military cadres.

Phase II Considerations

G-62. An increased emphasis on the study and evaluation of insurgent military tactics is required. Tactics during this phase are usually limited to ambushes, raids, sabotage, and terrorism. These activities provide the insurgent with supplies, experience, and self-confidence while eroding friendly morale and reducing friendly economic and military capabilities.

TRAINING

G-63. Insurgent training will be closely related to the tactics they use and will include vigorous political indoctrination. The combat forces and people within an area under the insurgents' political domination receive training. The insurgents carefully plan and train for individual operations and phases of movements.

Phase I Considerations

G-64. The insurgents train and indoctrinate their cadre and newly accepted or recruited indigenous personnel during this phase. Training consists of political indoctrination along with propaganda, communications, and intelligence collection techniques. Some training normally takes place in another country and a change in the number and type of personnel traveling to that country may indicate this fact.

Phase II Considerations

G-65. Intelligence analysts devote much attention to—

- Locating training camps and areas.
- Identifying training cadres.
- Interdicting the movement of insurgents to and from out-of-country training areas.

LOGISTICS

G-66. In an insurgency, as in conventional warfare, the insurgents' effectiveness is very much dependent on their logistical support. In an insurgency's early stages, the requirements for military equipment and supplies are less than in later stages. Accurate intelligence of the insurgents' sources and availability of supplies and equipment is essential to determine their capability to maintain and expand the insurgency.

Phase I Considerations

G-67. Two particular items have always been essential to the Phase I insurgents—money and a printing press. If they are highly successful in establishing and motivating their power base, they may never really have a need for the usual items of military supply. Money often comes from abroad, but bank robberies, unusual or excessive fund drives, payroll deduction requests, or sudden affluence among suspect government officials are cause for suspicion. Equipment to produce and disseminate propaganda (printing presses and radio sets) is of a special nature, and the friendly government can easily control its purchase and use.

Phase II Considerations

G-68. In this phase, logistics is a larger and more elaborate requirement for the insurgents. They must now get, store, transport, and maintain weapons, ammunition, explosives, signal equipment, and medical supplies. They now need more people to operate the logistics system. Insurgent supply caches or supply lines become critical concerns to friendly forces. Friendly forces must control the borders and coastlines. To detect or deter the movement of supplies, friendly forces must also use aerial surveillance over remote areas or areas the insurgents use.

EFFECTIVENESS

G-69. Effectiveness describes the qualitative ability of the insurgents to achieve their political or military purposes. The insurgents' effectiveness can be judged by the type and number of operations they are able to perform.

Phase I Considerations

G-70. In Phase I, the term "combat effectiveness" usually does not apply. Although the insurgents use words like "struggle" and "front," the words do not denote the use of armed forces. Usually there will be overt indications of the effectiveness of insurgent operations. Intelligence analysts may gather information on these operations through careful observation of organizations, movements, and elections at all levels. Penetration of these activities by government agents is very desirable and can make a significant contribution to the OB picture.

Phase II Considerations

G-71. The effectiveness factor now expands to include the combat efficiency of insurgent military forces. By carefully evaluating the other OB factors and taking note of actual combat experience, an intelligence analyst can evaluate the insurgents' combat effectiveness or lack thereof. The intelligence analyst can determine insurgents' strengths and weaknesses and, from this information, calculate their capability to follow various COAs.

PERSONALITIES

G-72. Personalities are not listed as a separate OB factor in a conventional situation. They are of greater importance in an insurgency, and as such, are listed as a separate factor.

Phase I Considerations

G-73. In Phase I, personalities are an extremely important factor. During this phase, when the insurgency is just beginning to organize and function and trying to spread its influence, the loss of a comparatively small number of men can practically destroy or set back its progress. The apprehension, compromise, or exposure of its leaders may destroy the insurgency. Knowing who the insurgent leaders are can also furnish a valuable indication of how insurgents train and how effective the overall effort will be.

Phase II Considerations

G-74. As in Phase I, personalities are important enough to warrant their consideration as a separate factor. Many insurgent units will use their commander's name rather than a conventional designation.

ELECTRONIC TECHNICAL DATA

G-75. In the early stages of OB intelligence, there is often a lack of uniform communications procedures. This fact prevents the development of an extensive electronic technical database. VHF citizens band sets may play a role in early terrorist operations. Equipment available to the insurgents will range from the most primitive to the most modern. Even equipment not generally available in the armed forces of major world powers, such as spread spectrum and frequency hoppers, can be easily obtained.

Phase I Considerations

G-76. The propaganda needs may result in insurgent-sponsored, medium-frequency or commercial radio AM broadcasts. Transmitters may be located outside national boundaries or in remote, inaccessible areas. These broadcasts frequently use code words to command and control insurgent operations. Later, there may be some increased use of VHF transmissions and more organized communications procedures. The standardization of communications practices reflects communications training.

Phase II Considerations

G-77. Much more extensive use of communications equipment characterizes this phase. Insurgents will capture equipment from government sources, purchase or steal it from commercial sources, have external sponsors who provide it, or obtain locally manufactured equipment. Communication procedures may reflect an external sponsor's doctrine and training practices.

MISCELLANEOUS

G-78. Other items contribute to knowledge of the insurgents, such as goals and methods. The following paragraphs discuss these items.

Phase I Considerations

G-79. This category can include—

- Historical studies of people and parties involved in the insurgency.
- Code names or numbers.
- Any other information that does not fit under the other eleven categories.

Phase II Considerations

G-80. Several miscellaneous items now become vital adjuncts to the other factors. Weapons, insignia, code names and numbers, and types and colors of uniforms and flags help identify enemy units. They also help identify the source of outside aid, the source of weapons and equipment smuggled into or purchased in the target nations, and the morale and effectiveness of the insurgent armed forces.

> *Note.* The following points must be remembered when applying the OB factors to an insurgency:
> - The insurgents' methods may change but their principles do not.
> - The previously discussed OB factors are closely interrelated and cannot be analyzed separately.
> - When an insurgency escalates to a Phase II situation, the OB effort must be expanded considerably. The enemy combat units must now be considered in addition to the various Phase I organizations and activities that will still be active.

SPECIAL INTELLIGENCE-GATHERING OPERATIONS

G-81. Alternative intelligence-gathering techniques and sources, such as doppelganger or pseudo operations, can be tried and used when it is hard to obtain information from the civilian populace. These pseudo units are usually made up of ex-guerrilla and/or security force personnel posing as insurgents. They circulate among the civilian populace and, in some cases, infiltrate guerrilla units to gather information on guerrilla movements and support infrastructure.

G-82. To persuade insurgents to switch allegiance and serve with the security forces requires much time and effort. Properly screened prospective candidates must choose between serving with the HN security forces and facing prosecution under HN law for terrorist crimes.

G-83. Government security force units and teams of varying size conduct infiltration operations against underground and guerrilla forces. They have been especially effective in getting information on underground security and communications systems, the nature and extent of civilian support and underground liaison, underground supply methods, and possible collusion between local government officials and the underground. Before such a unit can be properly trained and disguised, however, intelligence analysts must gather much information about the appearance, mannerisms, and security procedures of enemy units. Most of this information comes from defectors or reindoctrinated prisoners. Defectors also make excellent instructors and guides for an infiltrating unit. In using a disguised team, the selected men should be trained, oriented, and disguised to look and act like authentic underground or guerrilla units. In addition to acquiring valuable information, the infiltrating units can demoralize the insurgents to the extent that they become overly suspicious and distrustful of their own units.

INSURGENT COUNTERINTELLIGENCE AND SECURITY

G-84. Since COIN is the restoration of internal security in the AO, it demands a vigorous and coordinated COIN effort. Insurgents generate broad CI and security programs to thwart government penetrations. They set up security and early warning nets in rural and urban areas. These systems are composed of carefully recruited individuals chosen primarily because their work places them near sensitive insurgent installations. Typically, lookouts may be newspaper vendors, building janitors, young students, farmers, small shopkeepers, or fishermen.

G-85. These lookouts report possible government raids or other operations to liaison men chosen because they can travel without attracting notice. Liaison men are often letter carriers, taxi drivers, or traveling vendors who pass the information to insurgent officials.

G-86. The security and CI wing of the insurgent political organization produces false birth certificates, identification papers, and travel permits the agents require for travel, jobs, and other activities. To make it difficult for the police to check the authenticity of a forged document, the fictitious birthplace listed is often in a location that cannot be checked easily. Identity papers frequently list the bearer as a peddler, freelance writer, or artist because these occupations are difficult for the police to check. Insurgents sometimes avoid the forgery problem by stealing or buying genuine documents from some individual who they then may kill.

G-87. Meeting sites are a security problem. Insurgents prefer sites in which the arrival of several persons at about the same time will not attract attention or arouse suspicion. They favor woods and other secluded areas. When they must hold meetings at a house or apartment, they try to avoid those neighborhoods in which well-known antigovernment agitators live. Such areas may be under surveillance. They change meeting places frequently. When possible, they arrange meetings to coincide with some outwardly legal, proper reason for bringing individuals together. They stagger the arrivals and departures. Family members answer the door. Guards stay after the meeting to look for incriminating items left behind.

G-88. Insurgent groups routinely conduct security checks of members, potential members, and collaborators. Normally, they do not accept a recruit until they have investigated his present and past family, life, jobs, political activities, and close associates. A probationary period follows. If they urgently need a person with special skills, the insurgent group may bring in a person but assign him or her very limited tasks until the investigation is completed.

G-89. Insurgent groups test clandestine agents regularly. The insurgent security personnel may, without warning, summon an individual to test his reaction. If he is guilty of disloyalty, he may sense possible exposure and desert. Insurgent security personnel may keep a suspect ignorant of a change in meeting place. If government security forces show up at the original site, the insurgent organization knows the suspect is a government informant. Strict conformance with security procedures is required. Cell members are subject to punishment if they do not report violations. Security sections discover and liquidate hostile agents. They spend as much time, if not more, watching their own personnel as they do the enemy's.

FRIENDLY FORCES COUNTERINTELLIGENCE AND SECURITY

G-90. The techniques pertaining to friendly clandestine collection operations also apply to covert CI activities. The emphasis, however, is on information of CI interest rather than intelligence interest. But during CI operations, information of intelligence interest may also be obtained and should be passed to interested agencies.

G-91. Most of the CI measures used will be overt in nature and aimed at protecting installations, units, and information and detecting espionage, sabotage, and subversion. Examples of CI measures to use are—

- Background investigations and records checks of persons in sensitive positions and persons whose loyalty may be questionable.
- Maintenance of files on organizations, locations, and individuals of CI interest.
- Internal security inspections of installations and units.
- Control of civilian movement within government-controlled areas.
- Identification systems to minimize the chance of insurgents gaining access to installations or moving freely.
- Unannounced searches and raids on suspected meeting places.
- Censorship.

Glossary

AAR	after action review
AECA	Arms Export Control Act
AIAP	Army International Activities Program
AM	amplitude modulation
AO	area of operations
AOB	advanced operational base
AOR	area of responsibility
ARNG	Army National Guard
ARSOF	Army special operations forces
ARTEP	Army Training and Evaluation Program
BIIP	Bureau of International Information Programs
C2	command and control
CA	Civil Affairs
CAS	close air support
CAT	Civil Affairs team
CBJ	Congressional Budget Justification
CCIR	commander's critical information requirements
CD	counterdrug
CI	counterintelligence
CIA	Central Intelligence Agency
CIG	civil information grid
CIM	civil information management
CJCS	Chairman of the Joint Chiefs of Staff
CM&D	collection management and dissemination
CMO	civil-military operations
COA	course of action
COIN	counterinsurgency
COM	chief of mission
Comm	communications
COMSOC	Commander, Special Operations Command
CONPLAN	concept plan
CR	civil reconnaissance
CSDF	civilian self-defense force
CT	counterterrorism
CTFP	Counterterrorism Fellowship Program
DAO	Defense Attaché Office

DC	dislocated civilian
DFT	deployment for training
DNI	Director of National Intelligence
DOD	Department of Defense
DOS	Department of State
DSCA	Defense Security Cooperation Agency
DZ	drop zone
E&R	evasion and recovery
EOC	emergency operations center
EPA	evasion plan of action
ESF	Economic Support Fund
ETSS	extended training service specialist
EW	electronic warfare
FAA	Foreign Assistance Act
FHA	foreign humanitarian assistance
FID	foreign internal defense
FM	field manual; frequency modulation
FMF	Foreign Military Financing
FMS	foreign military sales
FNS	foreign nation support
FY	fiscal year
GCC	geographic combatant commander
GTA	graphic training aid
HA	humanitarian assistance
HCA	humanitarian and civic assistance
HMA	humanitarian mine action
HN	host nation
HQ	headquarters
HUMINT	human intelligence
IAW	in accordance with
ID	identification
IDAD	internal defense and development
IMET	international military education and training
IMINT	imagery intelligence
INL	International Narcotics and Law Enforcement Affairs
INTSUM	intelligence summary
IO	information operations
IPB	intelligence preparation of the battlefield
IPI	indigenous populations and institutions
IR	information requirement
ISR	intelligence, surveillance, and reconnaissance

JCET	joint combined exchange training
JCS	Joint Chiefs of Staff
JFC	joint force commander
JOPES	Joint Operation Planning and Execution System
JP	joint publication
JSCP	Joint Strategic Capabilities Plan
JSOTF	joint special operations task force
JSPS	Joint Strategic Planning System
JTF	joint task force
JWCA	joint warfighting capabilities assessment
KIA	killed in action
LOC	line of communications
LZ	landing zone
MASINT	measurement and signature intelligence
MCA	military civic action
MEDEVAC	medical evacuation
METL	mission-essential task list
METT-T	mission, enemy, terrain and weather, troops and support available—time available
METT-TC	mission, enemy, terrain and weather, troops and support available, time available, civil considerations
MI	military intelligence
MOS	military occupational specialty
MP	military police
MTP	mission training plan
MTT	mobile training team
NA	nation assistance
NADR	Nonproliferation, Antiterrorism, Demining, and Related
NCO	noncommissioned officer
NEO	noncombatant evacuation operation
NGO	nongovernmental organization
NLT	not later than
NMS	national military strategy
NSA	National Security Agency
NSC	National Security Council
NSS	National Security Strategy
O&M	operations and maintenance
OB	order of battle
OCONUS	outside the continental United States
OJCS	Office of the Joint Chiefs of Staff
OPCON	operational control

OPLAN	operation plan
OPORD	operation order
OPSEC	operations security
OSINT	open-source intelligence
OUSD(P)	Office of the Under Secretary of Defense (Policy)
PDSS	predeployment site survey
PEP	personnel exchange program
PIR	priority intelligence requirement
PKO	peacekeeping operations
PO	peace operations
POC	point of contact
POI	program of instruction
POL	petroleum, oils, and lubricants
POR	preparation of replacements for overseas movement
PPBS	Planning, Programming, and Budgeting System
PRC	populace and resources control
PSYOP	Psychological Operations
PW	prisoner of war
QRF	quick reaction force
ROE	rules of engagement
S-2	intelligence staff officer
S-3	operations staff officer
S-4	logistics officer
SA	security assistance
SAO	security assistance organization
SAT	security assistance team
SATMO	Security Assistance Training Management Organization
SCA	support to civil administration
SCG	Security Cooperation Guidance
SecDef	Secretary of Defense
SF	Special Forces
SFOD	Special Forces operational detachment
SFODA	Special Forces operational detachment A
SFODB	Special Forces operational detachment B
SFODC	Special Forces operational detachment C
SIGINT	signals intelligence
SIO	senior intelligence officer
SIR	specific information requirement
SITMAP	situation map
SOF	special operations forces
SOFA	status-of-forces agreement

SOP	standing operating procedure
SOTF	special operations task force
SSB	single sideband
TDY	temporary duty
TECHINT	technical intelligence
Trans	transportation
TSCP	theater security cooperation plan
TSOC	theater special operations command
TTP	tactics, techniques, and procedures
UCMJ	Uniform Code of Military Justice
UHF	ultrahigh frequency
U.S.	United States
USAID	United States Agency for International Development
USAJFKSWCS	United States Army John F. Kennedy Special Warfare Center and School
USAR	United States Army Reserve
USC	United States Code
USD(P)	Under Secretary of Defense for Policy
USDR	United States defense representative
USG	United States Government
USSF	United States Special Forces
USSOCOM	United States Special Operations Command
UW	unconventional warfare
VHF	very high frequency

This page intentionally left blank.

References

SOURCES USED:

These are the sources quoted or paraphrased in this publication.

AR 12-15, *Joint Security Assistance Training (JSAT)*, 5 June 2000

AR 600-8-101, *Personnel Processing (In-, Out-, Soldier Readiness, Mobilization, and Deployment Processing)*, 18 July 2003

FM 1-02, *Operational Terms and Graphics*, 21 September 2004

FM 3-0, *Operations*, 14 June 2001

FM 3-05.40, *Civil Affairs Operations*, 29 September 2006

FM 3-07.31, *Peace Operations: Multi-Service Tactics, Techniques, and Procedures for Conducting Peace Operations*, 26 October 2003

FM 3-13, *Information Operations: Doctrine, Tactics, Techniques, and Procedures*, 28 November 2003

FM 7-0, *Training the Force*, 22 October 2002

JP 3-07.1, *Joint Tactics, Techniques, and Procedures for Foreign Internal Defense (FID)*, 30 April 2004

JP 3-07.5, *Joint Tactics, Techniques, and Procedures for Noncombatant Evacuation Operations*, 30 September 1997

JP 3-57, *Joint Doctrine for Civil-Military Operations*, 8 February 2001

Title 10, United States Code, Section 375, *Restriction on Direct Participation by Military Personnel*, 3 January 2005

Title 10, United States Code, Section 401, *Humanitarian and Civic Assistance Provided in Conjunction With Military Operations*, 3 January 2005

Title 10, United States Code, Section 2011, *Special Operations Forces: Training With Friendly Foreign Forces*, 3 January 2005

Title 22, United States Code, Section 2151, *Congressional Findings and Declaration of Policy*, 3 January 2005

Title 31, United States Code, Section 1341, *Limitations on Expending and Obligating Amounts*, 3 January 2005

DOCUMENTS NEEDED:

These documents must be available to the intended users of this publication.

FM 3-05.102, *Army Special Operations Forces Intelligence*, 31 August 2001

FM 3-21.10, *The Infantry Rifle Company*, 27 July 2006

FM 4-01.011, *Unit Movement Operations*, 31 October 2002

FM 5-0, *Army Planning and Orders Production*, 20 January 2005

FM 34-130, *Intelligence Preparation of the Battlefield*, 8 July 1994

READINGS RECOMMENDED:

These sources include relevant supplemental information.

Bailey, Cecil E., "OPATT: The U.S. Army SF Advisers in El Salvador," *Special Warfare Magazine*, December 2004, pp 18-29

Defense Security Cooperation Agency, www.dsca.mil

FM 3-05.20, *(C) Special Forces Operations (U)*, 10 October 2006

FM 3-05.30, *Psychological Operations*, 15 April 2005

FM 3-05.201, *Special Forces Unconventional Warfare Operations*, 30 April 2003

FM 3-05.214, *(C) Special Forces Vehicle-Mounted Operations Tactics, Techniques, and Procedures (U)*, 10 October 2006

GTA 31-01-003, *Detachment Mission Planning Guide*, 1 March 2006

JP 1-02, *Department of Defense Dictionary of Military and Associated Terms*, 12 April 2001 (as amended [online] through 16 October 2006)

Index

By Order of the Secretary of the Army:

PETER J. SCHOOMAKER
General, United States Army
Chief of Staff

Official:

JOYCE E. MORROW
Administrative Assistant to the
Secretary of the Army

0701202

DISTRIBUTION:

Active Army, Army National Guard, and U. S. Army Reserve: To be distributed in accordance with initial distribution number 115096, requirements for FM 3-05.202.

PIN: 083806-000

www.ingramcontent.com/pod-product-compliance
Lightning Source LLC
Chambersburg PA
CBHW071721170526
45165CB00005B/2096